MATH
Connections®

A Secondary Mathematics Core Curriculum

1a

William P. Berlinghoff
Clifford Sloyer
Robert W. Hayden

IT's ABOUT
TIME®
HERFF JONES EDUCATION DIVISION

IT's ABOUT TIME ®

HERFF JONES EDUCATION DIVISION
84 Business Park Drive
Armonk, NY 10504
Phone (914)273-2233
Fax (914)273-2227
www.ITS-ABOUT-TIME.com

President
Tom Laster

Creative/Art Director
John Nordland

Director of Product Development
Barbara Zahm

Cover Illustration
Robert Conge

Illustrations
Dennis Falcon

Editorial Coordinator
Monica T. Rodriguez

Production Specialist
Bernardo Saravia

MATH *Connections®: A Secondary Mathematics Core Curriculum* was developed under the National Science Foundation Grant No. ESI-9255251 awarded to the Connecticut Business and Industry Association.

ISBN 1-58591-366-9
ISBN 1-58591-368-5 (Year 1, 2 Book Set)
'   4 5 6 VH 08 07 06 05

This project was supported, in part,
by the

National Science Foundation
Opinions expressed are those of the authors
and not necessarily those of the Foundation.

Welcome to **MATH** *Connections*®

This book was written for you. It is designed to provide you with mathematical experiences that will excite your curiosity, stimulate your imagination, and challenge your skills. It bridges mathematics with the real world of people, business and everyday life. It isn't finished until you take an active part in the interesting problems and projects that invite you to explore important mathematical ideas. You'll want to discuss these ideas with other students, your teacher and your family. You might find that not all your ideas work, but try again, perhaps a different approach will work—that is all part of learning. And the learning is up to you!

<u>In the Margins</u> **The Learning Outcomes** are in the margins of the first page of each section. These will alert you to the major topic. The **Thinking Tip** in the margins will help you in gathering your ideas and in solving problems. **About Words** will show you how some words we use in mathematics relate to words you already know and use every day. **About Symbols** will explain particular notations and their use in mathematics.

<u>In the Text</u> **A Word to Know** and **A Phrase to Know** appear in the text and signal particularly important definitions. Similarly, **A Fact to Know** signals an important mathematical result.

<u>In the Profiles</u> you will meet people in various careers and professions who use mathematics in their everyday work.

<u>In the Appendices</u> at the back of your book you'll find some more sections to assist with learning and problem solving.

• Appendix A: Using a TI-84 Plus (TI-83 Plus) Graphing Calculator

• Appendix B: Using a Spreadsheet

• Appendix C: Programming the TI-82 (T-83)

• Glossary

• Index

From time to time you'll see these graphic icons that call you to action.

Do this now
Identifies questions for you to answer.

Discuss this
Identifies questions for you to discuss as a class or in groups.

Write this
Usually requires you to gather information or reflect on a particular topic.

How MATH Connections takes you to the real world.

MATH Connections begins with you!
Each **MATH** *Connections* chapter introduces a concept by asking you to think about what you already know. You bring a lot of your life experiences into the classroom and with **MATH** *Connections* those experiences are strengths.

Provides a solid foundation in mathematics.
Building on your knowledge, **MATH** *Connections* connects your experiences with comprehensive mathematics. You'll learn algebra, geometry, statistics, probability, trigonometry, discrete mathematics plus dynamic programming, linear programming and optimization techniques.

Relates the mathematics to real situations.
As you learn the mathematics, you will apply it to real situations from hundreds of professions and careers ranging from architecture to micro-surgery to managing a grocery store. Whether it is at home, in games, in sports or at work .... **MATH** *Connections* connects mathematics to the real world of science, literature, art; and the things you do every day.

Think math. Do math. Talk math. Write math.
Ultimately, math is a language that can help you in every aspect of your life. And with **MATH** *Connections* you really make mathematics your own by exploring, looking for patterns and reasoning things out. Whether you are working on your own, in small groups or as a class to solve problems, with **MATH** *Connections*, you will achieve a real understanding of mathematics.

Classroom tested for excellence.
MATH *Connections* works! **MATH** *Connections* was field-tested by more than 5000 students like yourself, in more than 100 high school classrooms. During the 4-year field test, it was continuously refined by its developers and high school teachers. And year after year it has proven to make the learning of mathematics more effective and more enjoyable. Plus, bottom line, **MATH** *Connections* students score higher on state and national tests.

Prepares you for your future.
Whether you plan to pursue a career in the sciences, the fine arts, or sports .... **MATH** *Connections* prepares you for the real world and your future.

Algebra, geometry, probability, trigonometry, statistics, discrete mathematics, dynamic and linear programming, optimization...

MATH *Connections* ties these all together.
Just like bridge cables that sustain and connect the span of a bridge, **MATH** *Connections* supports and relates to what you do in school, at home, in games, in sports, in college and at work to make you stronger in math and stronger in life.

MATH *Connections* TEAM

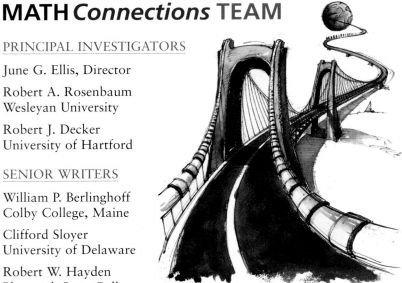

ADVISORY COUNCIL

James Aiello
Oakton Consulting Group
Oakton, Virginia

Laurie Boswell
Profile High School
New Hampshire

Glenn Cassis
Connecticut Pre-Engineering Program

Daniel Dolan
PIMMS, Wesleyan University
Connecticut

John Georges
Trinity College
Connecticut

Renee Henry
State Mathematics Consultant, Retired
Florida

James Hogan, Jr.
Connecticut Chapter of the National Technology Association

Lauren Weisberg Kaufman
CBIA Education Foundation

James Landwehr
AT&T Bell Laboratories
New Jersey

Donald P. LaSalle
Talcott Mountain Science Center
Connecticut

Daniel Lawler
Hartford Public Schools, Retired
Connecticut

Steven Leinwand
State Mathematics Consultant
Connecticut

Valerie Lewis
State Department of Higher Education
Connecticut

Gail Nordmoe
Cambridge Public Schools
Massachusetts

Thomas Romberg
University of Wisconsin

Kenneth Sherrick
Berlin High School
Connecticut

Albert P. Shulte
Oakland Schools
Michigan

Irvin Vance
Michigan State University

Cecilia Welna
University of Hartford
Connecticut

PRINCIPAL INVESTIGATORS

June G. Ellis, Director

Robert A. Rosenbaum
Wesleyan University

Robert J. Decker
University of Hartford

SENIOR WRITERS

William P. Berlinghoff
Colby College, Maine

Clifford Sloyer
University of Delaware

Robert W. Hayden
Plymouth State College,
New Hampshire

THE STAFF

Robert Gregorski
Associate Director

Lorna Rojan
Program Manager

Carolyn Mitchell
Administrative Assistant

THE CONTRIBUTORS

Don Hastings (retired)
Stratford Public Schools

Kathleen Bavelas,
Manchester Community-
Technical College

George Parker
E. O. Smith High School,
Storrs

Linda Raffles
Glastonbury High School

Joanna Shrader Panning
Middletown High School

Frank Corbo
Staples High School, Westport

Thomas Alena
Talcott Mountain Science Center,
Avon

William Casey
Bulkeley High School, Hartford

Sharon Heyman
Bulkeley High School, Hartford

Helen Knudson
Choate Rosemary Hall, Wallingford

Mary Jo Lane (retired)
Granby Memorial High School

Lori White Moroso
Beth Chana Academy for Girls,
Orange

John Pellino
Talcott Mountain Science Center,
Avon

Pedro Vasquez, Jr.
Multicultural Magnet School,
Bridgeport

Thomas Willmitch
Talcott Mountain Science Center,
Avon

Leslie Paoletti
Greenwich Public Schools

Robert Fallon (retired)
Bristol Eastern High School

Montville High School:
Lynn Grills
Janice Hardink
Diane Hupfer
Ronald Moore
Mark Popeleski
Henry Kopij
Walter Sherwin
Shari Zagarenski

Oxford Hills Comprehensive High School,
South Paris, ME:
Mary Bickford
Peter Bickford
Allen Gerry
Errol Libby
Bryan Morgan
Lisa Whitman

Parish Hill Regional High School:
Peter Andersen
Gary Hoyt
Vincent Sirignano
Deborah Whipple

Southington High School:
Eleanor Aleksinas
Susan Chandler
Helen Crowley
Nancy Garry
John Klopp
Elaine Mulhall
Stephen Victor
Bernadette Waite

Stonington High School:
Joyce Birtcher
Jill Hamel
Glenn Reid

TECHNOLOGY SUPPORT

Texas Instruments:
Graphing Calculators and
View Screens

Presto Press:
Desktop Publishing

Key Curriculum Press:
Geometer's Sketchpad

PROFILES

Barbara Zahm
David Bornstein
Alex Straus
Mimi Valiulis

PHOTO CREDITS:

Table of Contents

Thais Seldess
Surveying a Changing City

Thais Seldess is the field operations manager for Metro Chicago Information Center, a survey research company. Her job is to direct the collection of data in the field, then make sure that the data are sorted and organized for reports.

"There are a number of ways to collect data," she explains. "It depends on the kind of project. Often we do door-to-door studies. Or we can talk to people at malls. We might also use phone or mail surveys."

Thais was a psychology major at the University of Wisconsin. She also studied statistical analysis. Today, she uses everything she learned in school — and more. She designs questionnaires and makes up budgets for projects. And she's in charge of hiring and training all the people needed to collect data.

"In the Chicago area," Thais explains, "we have an ongoing Quality-of-Life Survey. Every year, we survey 3,000 random households. Making sure samples are random requires a knowledge of probability and possible outcomes. Then we take the raw data and translate it into a 'profile' of the city. This is used to spot trends in Chicago's population and highlight social problems.

"For example," she adds, "our data showed that in the early 1990s, more and more people were moving out of Chicago. But by the late 1990s, we saw people moving back. Museums and arts groups can use these numbers to improve services and attract more members.

"I like to see the data used in ways that improve people's lives. Then I feel I've accomplished something."

Turning Facts Into Ideas

CHAPTER 1

1.1 Data

- What is a fact?

- What is an idea?

- What's the difference between them?

- What do you suppose the title of this chapter is talking about?

The word **mathematics** comes from the ancient Greek language. In that language, it means "the things that are worth knowing." The first person to use it as a name for the study of numbers and shapes was Pythagoras, a Greek scholar and teacher who lived in the sixth century B.C. Since then, *mathematics* has come to mean a way of thinking about many things that are worth knowing.

These days, so much information comes at us from so many different sources—TV, radio, newspapers, books, magazines, computers—that our era is sometimes called "The Information Age." The part of mathematics called *statistics* gives us ways to organize and examine all that information, ways to make it work for us. This course begins with a look at some key ideas of statistics.

About Words

The noun *data* is plural; its singular form (not used very much) is *datum*, a single item of information. When *data* refers to a whole collection of pieces of information, it sometimes is used with a singular verb.

Pieces of information are called **data**. They are often, but not always, numbers. Your name, height, weight, birthday, and eye color are data. So are your school grades from last year, the sports scores in the newspaper, the listings in the TV program guide, and the food prices in your local supermarket.

a

Can you think of other kinds of data that you run across nearly every day? What kinds of data do you see at school? At home? In your neighborhood? In the newspaper? On TV?

Data are everywhere, sometimes just lying around for the taking. This abundance of data leads to two problems:

1. Before collecting data about a question, you must understand the question well enough to know which data to collect.

2. Once you have the data, they must be organized in ways that makes them easy to understand and use.

The first of these two problems can be harder than you might imagine. For instance, think about this question for a few minutes:

ARE BOYS TALLER THAN GIRLS?

The question seems simple enough, right? It just asks which are taller, boys or girls. But is the question really as simple as it appears? Let's think about how we might find a reliable answer for it.

Really understanding a question may require some careful thinking about the situation in which the question arises, and

CAREFUL THINKING TAKES TIME.

If a question is difficult for you, expect to think about it more than once before deciding what to do. Talking it over with someone else is a good way to think it through. Explaining a question to another person usually helps to focus your thoughts, and someone else's ideas can give you new ways of looking at yours.

b

Do you think boys are taller than girls? Whether your opinion is *yes* or *no*, how might you try to convince someone who disagrees with you? What data would you gather? How would you gather it? Collect some data.

After you understand a question and gather data about it, you need to organize the data. That is, you need to mold your bunch of separate bits of information into a single, useful tool for answering your question. This is the sense that justifies using data with a singular verb. But how do we do it? Look at the **Thinking Tip**. Can we break our problem down into smaller pieces that are easier to handle? Sure! Let's start with:

Thinking Tip
Ask a simpler question.
When you have a question that is too difficult to answer all at once, break it into smaller and smaller pieces until you find a piece that you think you can answer. Then build back up, piece by piece, to the original question.

HOW CAN DATA BE DISPLAYED CLEARLY?

That is, how can we show the data to someone in a simple, easy-to-understand way? For example, Display 1.1 shows the NCAA Division 1 women's basketball champions from 1983 to 2004. Is there a better way to present that data?

NCAA Women's Basketball Champions			
Year	Winning College	Year	Winning College
1983	USC	1994	North Carolina
1984	USC	1995	Connecticut
1985	Old Dominion	1996	Tennessee
1986	Texas	1997	Tennessee
1987	Tennessee	1998	Tennessee
1988	Louisiana Tech	1999	Purdue
1989	Tennessee	2000	Connecticut
1990	Stanford	2001	Notre Dame
1991	Tennessee	2002	Connecticut
1992	Stanford	2003	Connecticut
1993	Texas Tech	2004	Connecticut

Display 1.1

In Display 1.1 are 22 observations of *college* and *year*. The years provide one kind of data: numbers. The colleges that won are data in the form of words or labels. These data are not numbers, but we can use numbers to summarize them. Each winning college is a category, a particular kind of result (just as flavors represent the particular kinds of ice cream you might buy). We can count how many years fall into each of these categories. For instance, USC was the champion for two years, 1983 and 1984.

These questions refer to the data in Display 1.1.

1. How many different colleges have won the championship? That's the number of categories in this example.

2. Which college won the most championships?

3. List the winning colleges. Next to each one, write the number of years it has been champion.

Labels separate things into categories. Data that are labels or words will be called **category data** or **label data**. Data that are numbers being used to count or measure things will be called **measurement data**. The list of winning colleges in Display 1.1 is category data. Question 3 asked you to summarize this list in a shorter table by counting the years in each category.

The post office uses a lot of data about each package it delivers. Some is category data and some is measurement data.

1. Which of the following data about a package are category data, and which are measurement data?

   (a) the town (or city) in the address
   (b) the state in the address
   (c) its weight
   (d) the ZIP code
   (e) the postage on it

2. List at least three more items of data about a package. For each one, say whether it is category data or measurement data.

An effective way to display category data is by a picture that lists the different categories with a bar drawn next to each one. The number of times each label occurs in the data is shown by the length of the bar relative to some scale. Such a picture is called a **bar graph** (or **bar chart**).

Display 1.2 represents the data of Display 1.1. It is a typical example of a bar graph. The bars may be drawn horizontally or vertically.

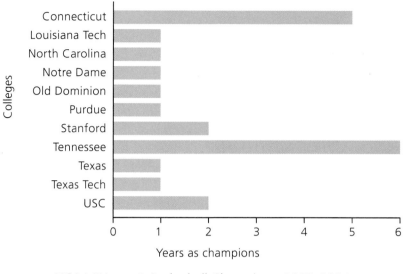

NCAA Women's Basketball Champions, 1983–2004

Display 1.2

Display 1.3 is another example of a bar graph. The data are the hair colors of the students in a ninth grade class. Use that bar graph to answer these questions.

1. What are the categories in this data set? How many are there?

2. Which hair color is most common in this class?

3. List the number of students with each hair color.

4. How many students are in the class? (Assume that each student has only one hair color!)

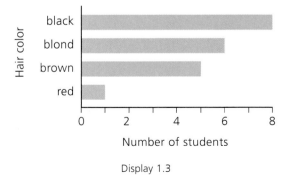

Display 1.3

Bar graphs can also be used for displaying measurement data. Some examples of that appear in the following problems.

Problem Set: 1.1

1. Display 1.4 shows the NCAA Division 1 men's baseball champions from 1970 to 2004.

   (a) How many different colleges won the championship during those years?

   (b) Which college won the most championships?

   (c) Make a table showing the number of years each college won the championship.

   (d) Display the information of your table in a bar graph.

   (e) At a quick glance, which do you think gives more information – your table or your bar graph?

NCAA Men's Baseball Champions, 1970–2004			
Year	Winning College	Year	Winning College
1970	USC	1988	Stanford
1971	USC	1989	Wichita State
1972	USC	1990	Georgia
1973	USC	1991	LSU
1974	USC	1992	Pepperdine
1975	Texas	1993	LSU
1976	Arizona	1994	Oklahoma
1977	Arizona State	1995	Cal. State, Fullerton
1978	USC	1996	LSU
1979	Cal. State, Fullerton	1997	LSU
1980	Arizona	1998	USC
1981	Arizona State	1999	Miami (FL)
1982	Miami (FL)	2000	LSU
1983	Texas	2001	Miami (FL)
1984	Cal. State, Fullerton	2002	Texas
1985	Miami (FL)	2003	Rice
1986	Arizona	2004	Cal. State, Fullerton
1987	Stanford		

Display 1.4

2. The bar graph in Display 1.5 shows the eye colors of students.

   (a) How many students are in the class?
   (b) How many students of each eye color are there?
   (c) How does this set of data compare with the data for your class?

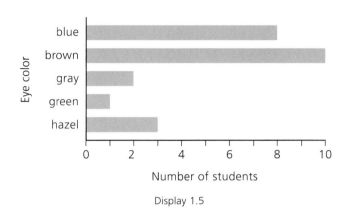

Display 1.5

For each set of data in problems 3, 4, and 5, draw a bar graph, if you can. Try to make your pictures clear and accurate. Make a note of anything that keeps you from being as accurate as you would like. After you finish drawing the pictures, answer these two questions for each data set.

   (a) At a quick glance, which gives more information—the table or the graph?

   (b) If you could use either a table or a graph, but not both, to display the data, which would you choose? Why?

3. In an effort to show how little money is devoted to its work, the National Aeronautics and Space Administration (NASA) compiled the data in Display 1.6. It shows how the 2000 federal budget was spent. Why do you think they gathered the data?

Category	Billions of $
Fixed Spending	1000
Defense/International	311
Domestic Programs	242
Interest on Debt	223
NASA	14

Display 1.6

4. Pilgrims Insurance Companies reported 2001 corporate travel expenses as a percentage of a total of $98 million, as shown in Display 1.7.

Category	%
Automobile	27%
Airfare	20%
Sundry	17%
Lodging	13%
Relocation	11%
Meals	7%
Entertainment	5%

Display 1.7

5. A survey of cigarette taxes per pack in developed countries where English is the main language yielded the data in Display 1.8. The amounts are expressed in 2002 U.S. dollars.[1]

Canada	$2.73
United Kingdom	$5.03
Ireland	$3.52
New Zealand	$2.89
Australia	$2.77
United States	$0.84

Display 1.8

6. Imagine if everyone were born at the same time of year! It doesn't happen that way, of course. On the other hand, "when you're born" is not evenly distributed over time, either. Display 1.9 shows the birth totals for each month of a year at a certain hospital.

(a) Represent these data by a bar graph.
(b) Is there a pattern in the data? Do they bunch? If so, where do they bunch?

[1]From Action on Smoking and Health (ASH) at www.no-smoking.org/july02/07-08-02-2.html. See ASH at http://ash.org

(c) Suppose someone asked you, "Which season during the year are humans typically born (winter, spring, summer, fall)?" What would you answer? What evidence do you have to support your answer? How might you gather data to support your answer? What data would you try to get?

Jan	Feb	Mar	Apr	May	June
20	26	32	37	29	20
July	Aug	Sept	Oct	Nov	Dec
18	15	12	13	14	17

Display 1.9

1.2 Finding the Mean

After studying this section, you will be able to:

Draw and interpret dotplots

Compute and describe the mean of a set of data

Determine the frequencies of values and the mode of a set of data.

You and your classmates have gathered some data about each other. Display 1.10 is a table of the kind of data you may have collected.

Name	Eye Color	Weight (lb)
Fred	blue	94
Marcia	blue	82
Bill	green	78
Pedro	brown	110
Annalee	blue	88
Sally	green	74
Li	brown	128
Max	orange	6
Cecile	brown	94
Alonzo	brown	126

Display 1.10

Look at the data in Display 1.10 and write a paragraph about what you see there. Here are some questions to get you started.

- Is one eye color much more common than the others?

- Is one eye color very rare?

- Are there any weights that are much larger than most? Much smaller?

Your paragraph should answer these questions. It should also include any other interesting things you notice.

About Words

The word *analyze* means to see what something is or how it works, by breaking it down and examining its parts. Its noun form is *analysis,* as in *handwriting analysis, chemical analysis,* and *psychoanalysis.* Its opposite is *synthesize,* which means to build something up from pieces.

A clear display is a first step toward making data work for us. But if we expect a set of data to be an effective tool, we have to do more than just see what it *says*. We have to analyze it carefully to understand what it *means*. How can we start to analyze a set of data? Go back to a **Thinking Tip**—ask some simpler questions. For instance.

Is there a typical data item or value?

If so, how close are the data to the typical one?

In some sense, the first question asks about the *center* of the data. These two questions focus our ideas better than the wide open invitation to analyze the data, but they aren't very precise yet. Nevertheless, they are good enough to help us with some examples.

For category data, the only center that makes any sense is the category or label that occurs most often. Sometimes this is called the **mode** of the data set. The number of different categories is a simple but crude way to summarize how a set of category data is spread out. For instance, the data in Display 1.1 are spread over eleven categories, and the mode is Tennessee.

Describe the center and spread for the eye color data of Display 1.10.

For measurement data, there are more precise ways of describing the center and how the data are spread out around it. For instance, look at this small set of measurement data, the scores of 12 students on a 10 point quiz:

$$3, 5, 7, 7, 7, 8, 8, 8, 8, 9, 10, 10$$

One way to picture these data is shown in Display 1.11. We stack up three boxes to represent the three 7s in our data, four boxes to represent the four 8s, and two boxes to represent the two 10s.

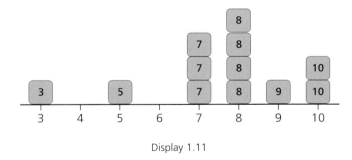

Display 1.11

Statisticians (and other people, too) use a method like the one shown in Display 1.11 to picture data sets, but they make it simpler. Instead of boxes with numbers, they use dots or some other easy-to-make marks to represent data items. Each data item is shown as a dot above its value on a number line. A diagram like that is called a **dotplot**. Display 1.12 shows a dotplot for the quiz scores.

Thinking Tip

Find or make examples. Often a problem is much easier to solve after you have seen a specific instance of it.

1.2 Finding the Mean

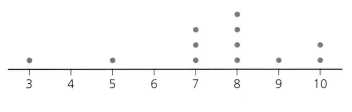

A Dotplot for the Quiz Scores

Display 1.12

Notice that some quiz scores occur more than once. The number of times a value appears in a data set is called the **frequency** of that value. In this case, 8 has frequency 4; 7 has frequency 3; 10 has frequency 2; and each of the other scores has frequency 1. If the frequency of some value is greater than every other frequency, that value is called the **mode** of the data. The mode of these quiz scores is 8.

- How do these quiz scores bunch together?

- Can you think of at least two different ways of describing the center of this set of scores?

- If you had to choose a single typical score, what would you choose? Why?

A common way to find a typical value for quiz scores is to add up the scores and then divide by the number of students who took the quiz.

> **A Word to Know:** The **mean** of a collection of measurement data is the sum of the data divided by the number of data items.

Find the mean of the quiz scores in Display 1.11.

Thinking Tip

Make a picture. A picture or diagram of a situation often leads to interesting ideas and helpful ways of thinking about problems.

One way to picture the mean is to think of it as a balance point for the set of scores. That is, if a 1 ounce weight were placed at each score on a weightless number line, then the mean would be at the point where the line balanced. (See Display 1.13.)

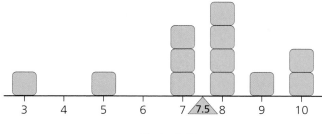

Display 1.13

Here are some easy ways to see how the mean acts as a balance point for data.

Equipment needed: A rigid 12-inch ruler (not a flimsy plastic one), a pencil, and five quarters or nickels or any coins that are all the same size.

- Put the pencil flat on a table or desk.

- Place the ruler across the pencil at its middle (the 6-inch mark), so that it balances.

1. Place one coin at the 3-inch mark, one at the 4-inch mark, and one at the 11 inch mark. (Note: When placing coins, try to center them on the marks.)

   (a) Does the ruler still balance? (It should.)

   (b) What is the mean of 3, 4, and 11?

2. Leave the coins at the 3-inch and 4-inch marks; remove the one at 11 inches. Keep the pencil at the 6-inch mark.

   (a) Find a way of placing *two* more coins on the ruler so that it balances.
   (b) At what numbers did you place the coins?
   (c) What is the mean of 3, 4, and the location numbers of the other two coins?
   (d) Can you find more than one way to place the other two coins so that the ruler balances? If so, how many can you find, and what are they? If not, why can't it be done?

These questions refer to the idea of the mean as a balance point.

1. In the example of Display 1.13, suppose the quiz grader made a mistake and each student lost one more point. Where would you expect the mean to be? Why?

2. Suppose you have just two 1-ounce weights on a weightless bar. Where is the balance point?

3. Suppose you have three 1-ounce weights and you know that two of them are at the same point on the number line and the other is six inches away, but you *don't* know exactly where they are on the line. What can you say about the location of the balance point?

4. Suppose you have a little sister who weighs only half as much as you do. If you take her to the seesaw at a playground, where would you have to sit to make the seesaw balance?

1. Make a dotplot of the weight data in Display 1.10.

2. What is the frequency of each value? Is there a mode? Why or why not?

3. Find the mean. Mark it on the number line of your dotplot.

4. In what way is the data for Max troublesome?

5. Ignore Max. He's the class mascot! Find the mean of the data without Max. Mark it on the number line of your dotplot.

6. In your opinion, which of these two means is a better measure of the center of the weight data for the class? Give a reason for your answer.

Problem Set: 1.2

In problems 1 through 3 below:

(a) List the frequency of each value. Find the mode, if there is one.

(b) Draw a dotplot to represent the data.

(c) Find the mean of the data set. Mark its location on your dotplot.

(d) Describe in words what the mean represents.

1. The school football team won all ten of its games this season. The winning margins for each game were

    7, 1, 13, 7, 6, 20, 3, 10, 7, 3

2. The data are the heights of you and your classmates, as measured and recorded in class.

3. Recent prices of half-gallons of different brands of milk in three supermarkets were

    $1.84, 1.81, 1.95, 1.88, 1.84, 1.85, 1.92, 1.88, 1.88, 1.81

4. Go to your local supermarket and write down the prices of 10 different kinds of dry breakfast cereal *in boxes with the same net weight.* Then find the mean price and the frequency of each different price.

5. This exercise refers to the data in Display 1.11. If you were the student who got the 9 on this quiz, then you had the third highest score in the class. Can you make up a set of quiz scores *with the same mean* that has your 9 as its highest score? If you can, do it. If you can't, write a short explanation of why you think it can't be done.

6. Make up two data sets that satisfy all these conditions:
   – Each data set contains 9 items.
   – Both data sets have the same mean.
   – Every value in the first set has frequency 1.
   – Every value in the second set has frequency 2 or 3.

7. Good Times Music, the local record store, keeps track of monthly sales of the Top 20 CDs by using a dotplot for each one. At the end of each day, the manager puts a dot on each CD's dotplot to show how many were sold that day. Display 1.14 shows the August dotplot for the latest Usher CD. There are only 27 entries because Good Times Music is closed on Sundays.

(a) Just by looking at the dotplot, estimate the mean daily total sales of this CD in August.

(b) Use the dotplot to list the 27 daily total sales for this CD.

(c) Find the mean of the data. Describe in words what this mean represents. How close did your estimate come to the actual mean?

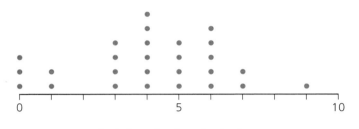

Daily Sales of Usher CD in August

Display 1.14

1.3 Displaying Data by Grouping

Suppose that a class of 20 students got these scores on a history test:

83, 94, 76, 85, 97, 88, 75, 86, 91, 51,
79, 86, 90, 81, 73, 79, 85, 86, 99, 93

How can we find the mean score? We could compute it by hand as before, by adding up the scores and dividing by 20. But there is less chance of making a routine arithmetic error if we use a calculator. Before doing that, it's a good idea to make a rough estimate of what the answer should be. In this case, notice that the scores tend to bunch up around 80 or a little more. This suggests that the mean should be not too far from 80.

Now let's work out the answer on the calculator. With a graphing calculator, you can do this in two ways:
- The old (probably familiar) way—add the numbers as you enter them, and then divide by 20.
- The new way—use the calculator's data handling tools.

The old way is easier to start, but it isn't as useful or reliable as the new way. The new way takes a couple of extra steps at the start, but from then on your job is easier. It stores each data number you enter in a separate place, allows you to look back over the list to make changes, and keeps the data to be worked on until you throw them away.

Enter the 20 history test scores shown at the beginning of this section into the first list of your graphing calculator.

To use the data handling power of the graphing calculator, start by entering each data item into one of the calculator's lists. (For details on how to do this, see Entering Data in a List in Appendix A.) At this point, the dull work is over. To find the mean, all you have to do is ask the calculator to display the one-variable statistics for that list. (See Summaries of 1-Variable Data in Appendix A for details on how to find and use this choice.) The display will include the mean, along with lots of other information about your data.

Learning Outcomes

After studying this section, you will be able to:

Use a graphing calculator to store a set of data, display it, and find its mean

Construct stem-and-leaf plots and histograms for data, and explain what these displays mean

Determine when stem-and-leaf plots and histograms are appropriate displays for data, and defend the decisions.

Thinking Tip

Estimate arithmetic answers. Little mistakes can lead to big errors. Even a calculator or a computer will give you garbage if you accidentally push the wrong button. Making a rough estimate of the answer often will help you catch those mistakes before they do any harm.

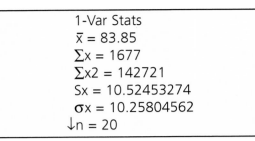

One Variable Statistics on a TI-82 (TI-83)

Display 1.15

Display 1.15 shows a TI-82 (TI-83) screen display of one variable statistics for the history test data you just entered. The symbols it uses are typical of many calculators (and many books, too), so it is a good idea to learn what they represent. The information related to the mean is on the second, third, and last lines of this display:

- The symbol $\bar{x}$ stands for the mean; it is read "x bar."
- The third line shows the sum of the data: $\sum x = 1677$.
- The bottom line shows that 20 data items are stored.
- The mean is $1677 \div 20$, which equals 83.85.

About Symbols

Σ is the capital Greek letter sigma, which corresponds to our capital S. It is commonly used in mathematics to stand for "sum."

Display 1.16 shows a dotplot of the history test scores balanced at the mean, 83.85.

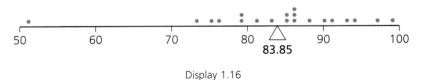

Display 1.16

 Use your graphing calculator to find the sum and the mean of the ten weights in Display 1.10.

When most of the items have frequency 1 or 2, as in Display 1.16, you may want to group the data in some way in order to display it clearly. When the data values are numbers—particularly two- or three-digit numbers—a common way to group them is by tens (because our numeration system is based on grouping by tens). That is, we group together all data values that either are the same or differ only in their final digit. In such cases, the grouping of the data is easy to see if it is organized in a form called a **stem-and-leaf plot**. Display 1.17 shows a stem-and-leaf plot for the 20 history test scores.

Stems	Leaves
5	1
6	
7	6 5 9 3 9
8	3 5 8 6 6 1 5 6
9	4 7 1 0 9 3

Display 1.17

A stem-and-leaf plot is a table of numbers separated by a vertical line. Each number on the left side (a stem) represents a tens group; the numbers to the right of it (the leaves) are the last digits of the data values in that tens group. There is a leaf for each data item, even if its value repeats an earlier one. For instance, look at the third line of the table:

7 | 6 5 9 3 9 stands for 76, 75, 79, 73, 79

In this way, the frequency of each tens group can be seen just by counting the leaves for its stem.

Do you think it is always (or sometimes or never) better to write the leaves of a stem-and-leaf plot in size order? Why? Here is a group exercise to help you think about the question.

1. Choose someone to be the official writer of a stem-and-leaf plot that you and your classmates are about to make.

2. Now, each person in the class chooses a year in the last hundred years. You can read it off a coin, or pick a favorite person's birth year, or just make one up.

3. The stem-and-leaf plot is to consist of all these dates. What should the stems be?

4. One by one tell your year to the official writer, who will chart it on the stem-and-leaf plot. Does the writer have to know all the dates before putting any of them in the chart, or can they be recorded as they are said?

5. Now order the leaves on each stem from smallest to largest, so that you have a second stem-and-leaf plot.

6. Finally, add the years 1945 and 1981 to your two stem-and-leaf plots. To which plot was it easier to add these dates?

Another handy feature of the stem-and-leaf plot is that it is easy to turn it into a *histogram*. Often, we do actually *turn* it 90° counterclockwise—so that the stems are on the bottom and the leaves make columns going up.

- The bottom line of the histogram shows the range of possible values of the data. For the stem-and-leaf plot of Display 1.17, the values are from 50 to 100. The bottom line is divided into sections of equal length.

- The vertical bars represent the frequencies of the data in each section. When turned up like this, the stems of the stem-and-leaf plot of Display 1.17 show the tens intervals on the bottom, and the leaves show the heights of the vertical bars.

Display 1.18 shows the resulting histogram.

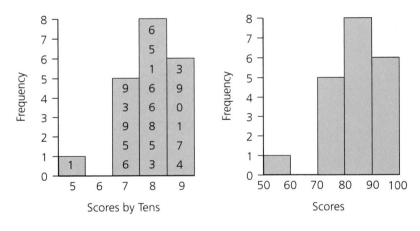

Display 1.18

 In what ways are histograms and bar graphs similar? In what important ways are they different? (*Hint:* Compare the histogram in Display 1.18 with the bar graph in Display 1.2 on page 7.)

A histogram is a picture that shows you in a general way how data are spread out and where they are bunched together. Making the size of the groupings (along the bottom line of the histogram) smaller gives the picture more detail; making the grouping size bigger gives less detail. Display 1.19 shows a histogram for the history exam scores grouped by fives, with the 0, 1, 2, 3, 4 leaves of each stem in one (narrower) bar and the 5, 6, 7, 8, 9 leaves in another. Display 1.20 shows a histogram for the same history exam scores grouped by twenties.

Can you make a stem-and-leaf plot for Display 1.19?
If you can, do it. If you can't, explain why you can't.

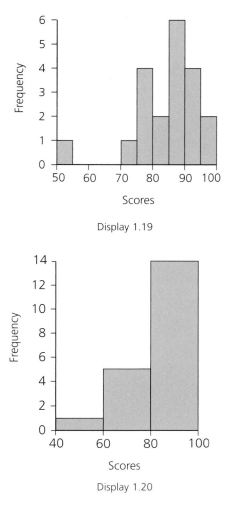

Display 1.19

Display 1.20

Each of the two histograms in Display 1.21 represents
a set of exam scores between 60 and 100, broken
down into 5-point score ranges.

1. How many scores does each histogram represent?

2. Which of these sets of scores has the higher mean?
   How do you know?

3. Construct a histogram for each one using the
   10-point score ranges (60–69, 70–79, etc.) and
   compare the two results. What do you notice?

Your graphing calculator can draw histograms. All it needs is the data and a few sizing instructions for the picture. The section of Appendix A titled *Drawing Histograms* contains detailed instructions for doing this.

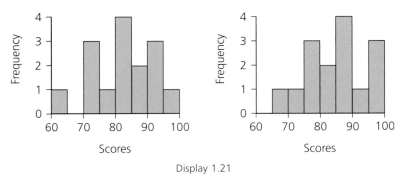

Display 1.21

Since the history exam scores from the beginning of this section are (or should be) still stored in your calculator, it is easy to get the machine to draw the histograms in Displays 1.18, 1.19, and 1.20. We'll help you with the first one; then you can do the other two on your own.

1. To draw the histogram in Display 1.18, begin by answering these questions:

   (a) How wide do you want each bar to be? Set the *x*-scale to that number.

   (b) At what *x*-value do you want the bars to start? Set the minimum for *x* to that number.

   (c) What was the highest possible score on this exam? Set maximum for *x* to that number.

   (d) What maximum and minimum settings should you choose for *y*? Why?

   (e) What scale setting should you choose for *y*? Why?

   (f) In which list is your data stored? How do you guide the calculator to that list?

   After you've chosen these settings, graph the result. Do you get the histogram in Display 1.18? You should. If not, go back and fix the settings until you do.

2. How must you change the settings in order to get the histogram in Display 1.19? Do it.

3. How must you change the settings in order to get the histogram in Display 1.20? Do it.

Reset the graph WINDOW settings of your calculator as they were for drawing Display 1.18. By changing *no more than five* of the original data values, can you get the calculator to draw the histogram shown in Display 1.22? Can you get the one in Display 1.23 in this way? If you can, do it. If not, explain why you can't. When you are finished with these questions, put back the original data. Save it for later use.

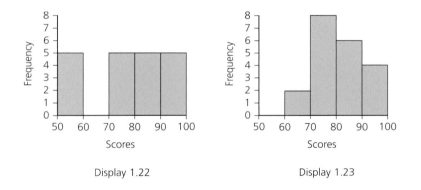

Display 1.22   Display 1.23

Problem Set: 1.3

1. (a) List the individual data items contained in the stem-and-leaf plot of Display 1.24. Then find the mean. Remember that: 2 | 3 1 ... stands for 23, 21, etc.

   (b) Make a histogram for this data set. In what way(s) is your histogram like the stem-and-leaf plot?

$$
\begin{array}{c|l}
2 & 3\ 1\ 5\ 0 \\
3 & 7\ 7\ 2\ 1\ 6 \\
4 & \\
5 & 3\ 9 \\
6 & 0\ 4\ 4\ 2\ 4\ 8 \\
7 & 2\ 1\ 2 \\
\end{array}
$$

Display 1.24

2. This exercise refers to Display 1.25, which lists the number of women in state legislatures in 2004.

(a) A stem-and-leaf plot for this data has been started in Display 1.26. Copy and finish it.

(b) Which data value has the highest frequency?

(c) Do you notice anything peculiar about the data? Explain.

(d) Which four states have the highest number of women in their state legislatures? Do these states have anything in common?

(e) Which three states have the lowest number of women in their state legislatures? Do these states have anything in common?

(f) Does this data set give you a good idea of how to compare women's political influence in different states? Why or why not? What other data would help you make a better comparison?

Women in State Legislatures					
Alabama	14	Louisiana	24	Ohio	26
Alaska	11	Maine	44	Oklahoma	22
Arizona	30	Maryland	64	Oregon	26
Arkansas	22	Massachusetts	49	Pennsylvania	32
California	37	Michigan	30	Rhode Island	19
Colorado	33	Minnesota	60	S. Carolina	15
Connecticut	54	Mississippi	22	S. Dakota	17
Delaware	21	Missouri	42	Tennessee	23
Florida	38	Montana	37	Texas	35
Georgia	43	Nebraska	12	Utah	21
Hawaii	21	Nevada	21	Vermont	60
Idaho	29	New Hampshire	130	Virginia	20
Illinois	49	New Jersey	19	Washington	49
Indiana	25	New Mexico	35	W. Virginia	22
Iowa	30	New York	50	Wisconsin	34
Kansas	53	N. Carolina	39	Wyoming	13
Kentucky	17	N. Dakota	23		

2004 National Conference of State Legislatures

Display 1.25

```
 0
 1
 2
 3
 4
 5
 6
 7
 8
 9
10
11
12
13
```

Display 1.26

3. The histogram in Display 1.27 shows the results of a 100-point biology exam.

   (a) How many students took the exam?
   (b) How many students scored between 70 and 79 (inclusive)?
   (c) How many students scored between 70 and 74 (inclusive)?
   (d) How many students scored between 80 and 89 (inclusive)?
   (e) How many students scored between 80 and 84 (inclusive)?
   (f) Can you say exactly what the scores were? Why or why not?
   (g) Find an approximate mean (class average) then explain how you did it.

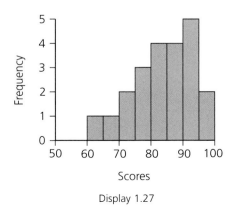

Display 1.27

4. The time of day when people are born is not evenly distributed over time. Display 1.28 shows the number of births for the hours of the day in one month at a certain hospital.

A.M.	12 – 1	1 – 2	2 – 3	3 – 4	4 – 5	5 – 6
	2	3	4	5	4	3
A.M.	6 – 7	7 – 8	8 – 9	9 – 10	10 – 11	11 – 12
	2	1	1	1	0	1
P.M.	12 – 1	1 – 2	2 – 3	3 – 4	4 – 5	5 – 6
	0	1	1	2	0	1
P.M.	6 – 7	7 – 8	8 – 9	9 – 10	10 – 11	11 – 12
	2	2	0	0	3	2

Display 1.28

(a) Represent these data by a histogram.
(b) In what way(s) are the data bunched (if at all)? Describe the typical time of birth in this hospital.
(c) Why would humans be born at one time of day, rather than at another?

For each of the data sets in problems 5 through 7:

(a) Estimate the sum and the mean of the data.
(b) Find the mean of the data set using a calculator to help with the arithmetic. Round to two decimal places.
(c) Check to see if your computed answer is close to your estimated answer. If not, look back to see which one is wrong and try again.
(d) Make a stem-and-leaf plot of the data.
(e) Draw a histogram of the data. To do this, you will have to decide how to group the data. Write a sentence or two explaining your choice of grouping. Then use your results to answer the specific questions in each problem.

5. To help plan its orders for frozen chicken, a restaurant does a two-week survey of the number of customers ordering chicken dinners. The daily totals are:

   47, 36, 55, 43, 24, 46, 38, 47, 52, 50, 38, 41, 45, 32

   (f) If frozen chicken costs the restaurant $1.25 for each order, what is the restaurant's average cost per day for chicken? How did you work out your answer?

   (g) If this problem were about fresh whitefish, would you answer the problem in the same way? Why or why not?

6. In *Consumer Reports*, the overall miles per gallon ratings of the 2002 small cars were as shown in Display 1.29.[1]

   (f) Which of these cars can be advertised honestly as having above average fuel economy for their class?

   (g) Is the mean that you computed a good estimate of the average miles per gallon of all the 2002 small cars of this type that are currently on the road? Why or why not?

Toyota Echo	38	Volkswagen Golf (4 cyl)	41
Chrysler PT Cruiser	18	Honda Civic	29
Mazda Protegé	27	Volkswagen Jetta (4 cyl)	24
Toyota Prius	41	Dodge Neon	23
Saturn S-Series	27	Hyundai Elantra	25
Nissan Sentra	26	Ford Focus	24

Display 1.29

[1]From *2002 Small Car Ratings*, ©2002 by Consumers Union of the U.S., Inc., Yonkers, NY 10703-1057, a nonprofit organization. Reprinted with permission from the April, 2002 issue of *CONSUMER REPORTS*® for educational purposes only. No commercial use of photocopying permitted. Log onto www.ConsumerReports.org.

7. Go into the school library and choose a shelf with at least 25 books on it. Write down the subject classification of that shelf. Then write down the exact number of pages of each of the first 25 books on the shelf: these numbers are your data. When you finish your calculations and drawings, compare your results with those of at least five other people in your class.

(f) Did you all get roughly similar results?

(g) If you did not, does the subject matter of the books seem to relate to the differences in some way? Do you think there might be some sort of pattern or predictable relationship between subject matter and book length? If so, what do you think it might be?

(h) How might you gather data to test your answers in part (g)?

1.4 Another Center: The Median

The mean is the most common way of describing the center of a set of measurement data, but it is not always the best. For instance:

Five students take a 100-point makeup exam. Four of them study hard, and each scores 90%. One doesn't study at all and only scores 10%.

1. Make a dotplot for these scores.

2. What is the mean score? Mark it on your dotplot.

3. Is the mean a good measure of center for these scores? Is it a typical value? Explain.

4. Can you find a better measure of center for these scores?

5. Would your answers to parts 3 and 4 change if the scores were 91, 92, 93, 94, and 10? Explain.

Now think about this example:

A big city car dealer wants to attract new customers. His ad reads:

> **See the new cars at Ari's Auto Palace!**
> **—First-come, first-served raffle—**
> **only 1000 $1 tickets to be sold**
> ***** 100 winners will be drawn *****
> **$599 in cash prizes!!!**
> **YOU have 1 chance in 10 to be a winner!**

Without knowing the amounts of the various prizes, can the mean value of a prize be determined? If so, what is it? If not, explain why it cannot be done.

Learning Outcomes

After studying this section, you will be able to:

Compute and interpret the median of a set of data

Describe how outliers affect the mean and the median

Determine the most appropriate measure of center for a set of data and justify the choice.

Is that a pretty good deal? If you bought a ticket, would you have about 1 chance in 10 of winning $5.99? Maybe, maybe not. Display 1.30 shows three possible raffle setups, each with 100 winners:

Raffle 1:	• 100 prizes of $5.99
Raffle 2:	• 1 prize of $500
	• 1 prize of $50
	• 98 prizes of 50 cents
Raffle 3:	• 1 prize of $100
	• 2 prizes of $50
	• 22 prizes of $10
	• 26 prizes of $5
	• 49 prizes of $1

Display 1.30

 Find the mean prize value for each of the raffles in Display 1.30. What do you notice about the three means? Is it what you expected? Why or why not?

The answers to these questions should show that the mean may not always be the best measure of the center of data. Only in Raffle 1 does the mean represent the value of the prize a winner is most likely to get.

In Raffle 2, the mean is not a good measure of center, because a few data numbers (two, in this case) are far away from the rest. The mode (50 cents) is much better than the mean at predicting what you are most likely to win—*if* you win at all—because 98 prizes out of 100 are worth only 50 cents each.

In Raffle 3, however, the mode doesn't work well as the typical prize. The $1 prize is the mode because it has the highest frequency (49), but *51 of the prizes are worth at least $5*. This means that a winner is more likely to win at least $5 than to win only $1. In other words, $5—which is in the middle of the data when they are arranged in size order—is a more typical prize in this case.

1. Which of these three raffles is the best deal? Why?

2. Would Ari make or lose money on these raffles? How much?

3. Do you think any one of them is worth buying a ticket for? Why or why not?

a

In Raffles 2 and 3, the top two prize values are much larger than the rest. When these values are averaged in, they cause the mean (the balance point) to be higher than a typical prize value. These values are called *outliers*. In general, an **outlier** is a value that does not "fit in well" with the rest of the data. Often it is much larger or smaller than most of the data. When a data set contains outliers, the mean may not be a good measure of center.

About Words

In geology, an *outlier* is a part of a rock formation that has been separated from the main formation by erosion.

1. Explain why the mean is often not a good measure of center for a data set with outliers. (*Hint:* In Raffle 2, if you throw out the two outliers, what is the mean of the rest of the prizes?)

2. Can you think of any cases in which the mean is not affected much by the presence of outliers?

b

Here's another example. You are the mayor of a village with eight small businesses and a shoe factory. The Census Bureau asks you how many people are employed by a typical company in your village. You do a quick phone survey and get these numbers:

1, 1, 2, 3, 4, 5, 6, 7, 79

What number do you tell the Census Bureau? Would the mean, 12, tell them what they wanted to know? Would the mode, 1, tell them what they wanted to know? Doesn't it make good common sense to choose the middle number, 4, to represent the number of employees of a typical company in your village?

"OK," we hear you say, "but what if there were only eight companies altogether? What if the data were

1, 2, 3, 4, 5, 6, 7, 79

Then there is no middle number."

That's a good point. What's a reasonable way to get around that problem? The main idea here is middle, so how about splitting the difference between the two numbers closest to the middle? In this case, those two numbers are 4 and 5, so you could use 4.5.

Would you tell the Census Bureau that a typical company in your village employs four *and a half* people? What does the number 4.5 represent in this situation? How would you phrase your report?

About Words

The *median* of a divided highway is the center strip that separates traffic in one direction from traffic in the other.

These examples illustrate another measure of the center of data.

A Word to Know: The **median** of a set of numerical data is the middle value when the data are arranged in size order. If there is an even number of data items, the median is the mean of the two middle ones.

In Raffle 3, the median is the mean (the average) of the 50th and 51st prizes, ordered by size. Since both of them are $5, the median is $5. Display 1.31 shows the three different measures of center for the cash prizes in each of the three raffles.

	Mean	Median	Mode
Raffle 1	$5.99	$5.99	$5.99
Raffle 2	$5.99	$0.50	$0.50
Raffle 3	$5.99	$5.00	$1.00

Display 1.31

Using only numbers between 1 and 10 (inclusive), choose a set of numbers with a mean that is larger than the median by at least 3. Then choose another set of numbers with a mean that is *smaller* than the median by at least 3. Try to do these things using as few numbers as you can.

Computing the mean, median, and mode by hand can be tedious. Having a calculator to add and divide makes it easier to find the mean, but the median and the mode can still be annoying if the data are not arranged nicely. For instance, consider this set of retail prices (in dollars) for compact disk players.

150, 100, 120, 140, 145, 100, 150, 80, 200, 80, 75, 155, 200, 250, 200, 160, 270, 165, 240, 290, 350, 140

It is not obvious from looking at these numbers whether there is a most common price, or what the middle number is. Fortunately, graphing calculators will do most of the annoying work for us if we know how to ask them.

1. Can you estimate the mean of these compact disk player prices? Try it. What about the median? Can you estimate it without putting all the numbers in size order? Can you find the mode? Write down your estimates now and save them to compare with the calculator's answers.

2. Use your graphing calculator to find the mean, median, and mode of the CD player prices. (Refer to Appendix A if you need detailed instructions about how to do this.)

3. When you have found the mean, median, and mode, compare them with the estimates you made. How close were you? If you were not very close, why do you think that happened? Do you see anything that might help you make better estimates the next time you have a problem like this?

4. Have the calculator draw a histogram of these prices. To do this, you will have to set the WINDOW values in some reasonable way. Choose values that you think will make your histogram show the information clearly. Then explain briefly why you chose these values.

After you have worked through the CD player example and still have the data in your calculator, try answering these questions:

Suppose that a new CD player appears on the market—a limited edition, Elvis Presley commemorative model, complete with gold inlay designs and platinum buttons that sells for $1,995. Put this new piece of data with what you already have.

1. What is the new mean price? By how much does it differ from the old mean?

2. What is the new median price? By how much does it differ from the old median?

3. If you were shopping for a new CD player, would it be more helpful to know the mean price or the median price? Why?

Problem Set: 1.4

You may use a graphing calculator for any of these problems. They may also be done without a calculator (but you'll need some patience).

1. Use the data in Display 1.32 to answer these questions.

   (a) Compute the mean number of moons.
   (b) Compute the median number of moons.
   (c) Find the mode for the number of moons, if there is one.
   (d) Make a dotplot of the data. Mark the mean and the median on the number line of your dotplot.
   (e) Are you comfortable using the mean to describe the average of the data? Why or why not?
   (f) Are you comfortable using the median to describe the average of the data? Why or why not?
   (g) Why is there such a difference between the mean and the median of these data?

Planet	Number of Moons
Mercury	0
Venus	0
Earth	1
Mars	2
Jupiter	16
Saturn	30
Uranus	20
Neptune	8
Pluto	1

Display 1.32

2. Find the mode of the eye colors of the people in your class. Does it make sense to ask for the mean or the median of these data? Why or why not?

3. A class of 15 students recorded the total number of hours they watched television during a six-week period. Here is the data they collected:

78, 82, 95, 85, 73, 99, 95, 89, 90, 86, 23, 92, 80, 84, 85

   (a) What are the mean and the median of these data? What does each number represent?

   (b) Make a stem-and-leaf plot of these data. Put a circle around the digit that represents the median.

   (c) The class learned that the student who watched TV for only 23 hours during the six week period had a broken television for part of that time. They decided to omit that number and analyze the data again. How much did this change the class mean? How much did it change the class median? Do you think it was a good idea for the students to omit that piece of data? Why or why not?

4. This problem refers to Display 1.25 which lists the number of women in each state legislature.

   (a) Find the mean, median, and mode of this set of data.

   (b) If New Hampshire were not included in the list, how would the mean, median, and mode be affected? Calculate these three measures of center again—without New Hampshire—and discuss the changes you discover.

5. Suppose that you want to show a friend how changing one or two data values can affect the mean or the median.

(a) Make up a set of 12 numbers between 0 and 10 (inclusive) with these properties:
   – No number appears more than three times.
   – The mean is 7.
   – The median is also 7.

(b) By changing no more than two numbers in your set, make it come out with the same mean, 7, but with median 8. You may have to start with a different solution for part (a) to make this work.

(c) By changing no more than two numbers in your original set, make it come out with the same median, 7, but with mean 6. You may have to start with different solutions for parts (a) and (b) to make this work.

6. To analyze the balance of answers for 5000 *true/false* questions in its data bank, a standardized testing service assigned the value 1 to each question with answer *true* and 0 to each question with answer *false*. They then computed the mean and the median.

(a) The median came out to be 1. What does this tell you about the number of *true* answers and the number of *false* answers? Why?

(b) The mean came out to be 0.7. What does this tell you about the number of *true* answers and the number of *false* answers? Why?

(c) Which number tells you more in this situation, the mean or the median? Why?

(d) Suppose, instead, that the mean came out to be 0.45. What would the median be? Explain.

1.5 Boxplots

Two boatloads of tourists went sport fishing. There were 11 tourists in each boat. In the first boat, every tourist caught three fish. In the second boat, five tourists caught six fish each, one caught three fish, and five caught none. We can think of these numbers of fish caught as two 11 number data sets, one for each boat.

1. What is the mean and the median of each data set?

2. Make a dotplot for each data set.

3. How would you describe the difference between these two sets of data?

As you can see from the "fishy" data in the questions above, the mean and the median do not tell us much about how far the data are from these centers or from each other.

Here is another example. These three data sets

(1)   1, 1, 4, 6, 9, 13, 15
(2)   2, 5, 6, 6, 7, 7, 16
(3)   2, 2, 2, 6, 10, 11, 16

have exactly the same mean, 7, and median, 6, but they are spread out along the number line very differently, as you can see from Display 1.33.

Suppose that data sets (1), (2), and (3) were the normal daily high temperatures, measured in degrees Celsius, for three different cities during one week of winter vacation. If you could go to any one of the three, which would you pick? Why?

Learning Outcomes

After studying this section, you will be able to:

Compute and interpret the five-number summary of a data set

Construct a boxplot and explain how it represents a set of data

Compare and contrast the data represented by two or more boxplots.

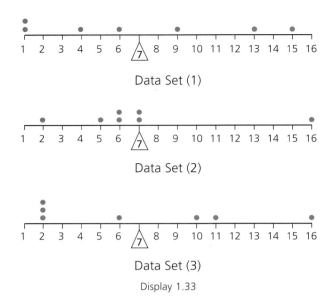

Display 1.33

Knowing something about how data are spread out is as important as knowing the mean and the median. For instance:

 Suppose you are going on a vacation trip to a place where the average daily temperature is 75° Fahrenheit. What clothes would you pack? Would it make a difference if you knew that the average came from the fact that it hits a low of 35° at night and a high of 115° in the daytime? Would you pack differently if you knew that the average daily temperature came from a low of 65° and a high of 85°?

We want to find a number or a very few numbers that measure how the data are spread out. The simplest measure of this is the range, the difference between the largest and smallest values.

 Data sets (1), (2), and (3) of Display 1.33 all have the same range. What is it?

The three data sets of Display 1.33 have the same mean and the same range, but they are spread out very differently, as you can see. Thus, the range is not a very precise way to describe the spread of a data set.

A somewhat better way is to use something called a boxplot. Here's an example of how it works:

Your school basketball team is getting ready for a statewide championship game. The other team, the Aces, comes from a small town far across the state from you and you don't know much about them. At first, all you can find out is that they scored 78 points in their best game of the season, and that their worst game score was a dismal 33 points. This tells you the range of their season's scores (what is it?), but it doesn't tell you how many points they score most of the time. Display 1.34 shows the range of scores as a horizontal line above the scale.

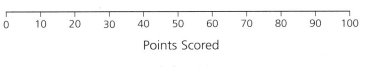

Display 1.34

A team member says that his mother overheard somebody talking about the Aces' record. The only thing she can recall hearing is that their median score for the season was 56. That helps a little. At least, now you know that in half their games they scored no more than 56 points. Display 1.35 shows the median score as a short vertical line crossing the range of scores.

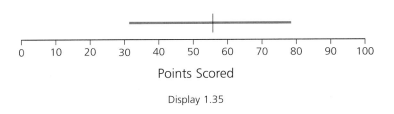

Display 1.35

Finally, the newspaper publishes a scoring summary for all the teams in the state. Instead of giving the points scored for every game, it shows a *five-number summary* for each team.

A **five-number summary** shows the high, low, and median numbers of the data, and it also shows two other numbers.

About Words

The suffix *-ile* means like or related to. Thus, a *quartile* is related to the idea of quartering–dividing something into four equal parts.

These two numbers and the median are called "quartiles." Just as the median divides the data in half, the quartiles divide the data into quarters. That is, if you list the data in order:

- one-quarter of the data comes before the first quartile;
- half (two-quarters) of it comes before the median (second quartile); and
- three-quarters of it comes before the third quartile (provided that you don't count any data points that are themselves quartiles).

Soon we'll tell you more precisely how to find the quartiles of a set of data. For now, here are two examples to help you remember that this is just the commonsense idea of dividing (approximately) a data set into quarters:

The first, second, and third quartiles of this 15 element data set are underlined:

1, 2, 3, <u>4</u>, 5, 6, 7, <u>8</u>, 9, 10, 11, <u>12</u>, 13, 14, 15

Each quartile of the following eight element data set is halfway between two data values. They appear here in parentheses:

1, 2, | 3, 4, | 5, 6, | 7, 8
(2.5)　　(4.5)　　(6.5)

 Back to the Aces' season. The first quartile is 49, and the third quartile is 66 for the Aces' scoring. Combine this information with what you already know to write a five-number summary of the Aces' scoring this season.

Knowing that the first quartile is 49 and the third quartile is 66 tells you that the Aces' scores for half of their games are between 49 and 66. This middle half gives you a good idea of how well the Aces usually play, so in one sense this part of the data is more important. We show this added importance in our picture by making that part of the line fatter; in fact, we turn it into a box, as shown in Display 1.36.

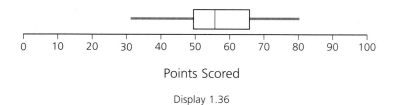

Points Scored

Display 1.36

Suppose that the Aces played 20 games during their season. Make up a set of scores that fits their five-number summary.

The diagram in Display 1.36 is called a *boxplot.* Sometimes it is called a *box-and-whisker diagram.* A **boxplot** is a five-number summary in picture form. The box shows where the middle 50% of the data lie. The difference between the upper and lower ends of this middle half is called the **interquartile range**. The horizontal lines on each end of the box (the whiskers) show the ranges of the lowest and highest quarters of the data. A vertical line across the box shows where the median is.

To make a boxplot, all you need is a five-number summary. Here's a step-by-step way to get a five-number summary from a set of data.

1. Arrange the data in size order, from smallest to largest.

2. Find the median. This is the *second quartile.*

3. Look at the set of numbers *before* the median and find its median. This is the *first quartile.*

4. Look at the set of numbers *after* the median and find its median. This is the *third quartile.*

5. Write down the five-number summary: smallest number, first quartile, median, third quartile, largest number.

Now make a boxplot by plotting these five numbers on a number line and drawing the box and whiskers, as in Display 1.36.

Let's look back at the three small data sets at the beginning of this section. They are almost too small to bother with, but they provide some easy practice making boxplots. Data set (1), already in size order, is

$$1, 1, 4, 6, 9, 13, 15$$

Its median is 6. The first quartile – the median of 1, 1, 4—is 1. The third quartile the median of 9, 13, 15—is 13. Thus, the five-number summary for data set (1) is

$$1, 1, 6, 13, 15$$

a

Find the five-number summaries for the temperature data sets (2) and (3) at the beginning of this section. Then compare your results with the boxplots for all three sets, which are shown in Display 1.37. Notice how the boxplots show differences in the spreads of these three sets.

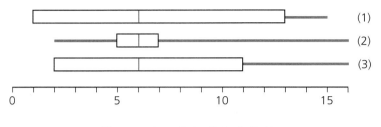

Temperature Data Sets 1, 2, 3

Display 1.37

Make a boxplot of your basketball team's game scores from last season. Then compare it with Display 1.36 to see how your team matches up with the Aces. If your team played the Aces, who do you think would win? Write a brief explanation of your answer.

You can get your graphing calculator to draw boxplots. All it needs is the data and a few sizing instructions. The section titled Drawing Boxplots in Appendix A explains how to do it.

b

The 20 history test scores from the beginning of Section 1.3, p.19, should still be stored in your graphing calculator. If they aren't, put them in again.

1. Use your calculator to draw a boxplot of these data. (Use Appendix A to help you.)

2. Delete the lowest test score from your data list and have the calculator draw the boxplot again, using the same WINDOW settings.

3. What is the main difference between these two boxplots? Which do you think is a better illustration of the data?

Problem Set: 1.5

You may use a calculator for these problems whenever it helps you.

1. Here is a five-number summary of a set of 40 test scores. All the test scores are whole numbers.

    $$53, 65.5, 78.5, 88.5, 97$$

    (a) Make a boxplot for this set of data.
    (b) In which quarter of the data set is 82? Mark this point (approximately) on your boxplot.
    (c) How many scores can you *guarantee* to be less than 82? Do you think there may be more? Why or why not?
    (d) How many data items lie between 65.5 and 78.5?
    (e) How many data items are less than 89?

2. Display 1.38 shows boxplots of runs scored in each game of the season for two Texas baseball teams, the Armadillos (A) and the Broncos (B).

    (a) Write a five-number summary for each one.
    (b) Which team scored at least 6 runs in half of its games?
    (c) Which team scored at least 9 runs in one-fourth of its games?
    (d) If these two teams played each other, which do you think would win? Why?

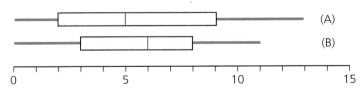

Display 1.38

3. You have finally saved enough money to buy that new mountain bike you've had your eye on for a long time. In *Consumer Reports*, you find the data shown in Display 1.39.

   (a) Using your calculator, enter the data. Then write the five-number summary and make a boxplot.

   (b) Notice that the bikes are listed in order of estimated quality within each category. Keeping this in mind, use the five-number summary to defend or reject the choice of the Specialized Rockhopper as one of *Consumer Reports'* 'quick picks' in the front-suspension category. Can you similarly defend their choice of the Specialized FSR XC Pro in the full-suspension category?

   (c) If you use only those bikes labeled as 'quick picks,' how does this affect your answer to part (b)?

Brand and Model	Price
Full-suspension models	
* Cannondale Jekyll 400	$1,100
* Specialized FSR XC Pro	$1,200
* Trek Fuel 80	$1,150
* Giant NRS 2	$1,450
Gary Fisher Sugar +4	$1,100
Front-suspension models	
* Gary Fisher Marlin	$450
* Specialized Rockhopper	$440
* Trek 4500	$440
* Raleigh M60	$450
Giant Yukon	$440
Schwinn Mesa GSD	$500
GT Avalanche 2.0 Disc	$500
Mongoose Rockadile ALD	$400
Hybrid models	
* Giant Cypress LX	$470
* Jamis Tangier	$480
* Gary Fisher Nirvana	$470
Specialized Crossroads Elite	$425
Comfort models	
* Jamis Explorer 2.0	$300
* GT Timberline	$300
* Trek Navigator 100	$300
* Schwinn Sierra GS	$300
Specialized Expedition	$280
Raleigh SC30	$290
Diamondback Wildwood	$285

Note: The items with an asterisk are *Consumer Reports'* Quick Picks. Within each category, the bikes are listed in order of best performance.[2]

Display 1.39

---

[2] From *Bicycles: More bike for the buck*, © 2004, by Consumers Union of the U.S., Inc., Yonkers, NY 10703-1057, a nonprofit organization. Reprinted with permission from the July, 2004 issue of *CONSUMER REPORTS*® for educational purposes only. No commercial use of photocopying permitted. Log onto www.consumerreports.org.

4. These questions refer to Display 1.40.

(a) About how much did the most expensive of these CD players cost in 2002?

(b) What was the approximate median price for these CD players in 2002?

(c) How many of these CD players cost more than the median?

(d) If you had saved $206 (the third quartile) to buy a CD player, how many of these models could you afford to choose from?

(e) What percentage of these CD players cost more than $147 (the first quartile)?

(f) Why isn't the median line in the middle of the box?

(g) Why is one whisker longer than the other?

(h) A January 2003 *Consumer Reports* article claimed that Philips MC-50 (at $150) had the highest overall score, followed by Panasonic SC-AK300 (at $170), Rio EX1000 (at $275), Panasonic SC-PM12 (at $170), and Panasonic SC-PM07 (at $150). If you were writing the article for *Consumer Reports*, would you label any of the top five as Best Buys? Defend your answer.[3]

[3] From *Minisystems*, © 1999 by Consumers Union of the U.S., Inc., Yonkers, NY 10703-1057, a nonprofit organization. Reprinted with permission from the January, 2003 issue of *CONSUMER REPORTS*® for educational purposes only. No commercial use of photocopying permitted. Log onto www.consumerreports.org.

(i) Write a short paragraph describing in words all the price information contained in Display 1.40.

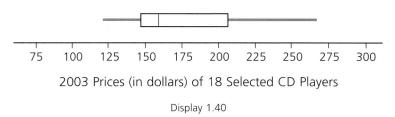

2003 Prices (in dollars) of 18 Selected CD Players

Display 1.40

5. Canada geese spend their winters in Connecticut. It may sound silly because Connecticut, a state in the Northeast, is cold in winter but compared to their home in Canada, Connecticut winters are quite warm! You've been studying the numbers of goslings they had in their gaggles in the past spring, and you find these numbers in the family units:

$$7, 9, 12, 9, 13, 8, 7, 6, 13, 7, 9, 13$$
$$8, 11, 14, 10, 8, 15, 14, 12, 6, 9, 13, 10$$

(a) What's a *gosling*? What's a *gaggle*?
(b) Make a dotplot of the data.
(c) What is the typical size of a gaggle of goslings (ignoring the adults)?
(d) Draw a boxplot for the data.
(e) What can you see in the dotplot that you can't see in the boxplot?

6. This problem is related to problem 5. Display 1.41 shows boxplots of that population of geese gaggles from earlier years.

   (a) In which of those years did the largest median gaggle size occur?

   (b) In which of those years was the largest gaggle born?

   (c) In which of those years were there gaggles that had no goslings?

   (d) In which of those years did the gaggles show the widest variation in size?

   (e) Write a short paragraph comparing your boxplot from problem 5 with the ones in Display 1.41. Answer these questions in your paragraph.

      (i) Which of the five earlier years has data most like your data? Which one appears to be most different? Justify your choices in terms of the ranges, the medians, and the interquartile ranges.

      (ii) Point out at least two clear differences between your set of data and that of the year you chose to be *most like* yours. Do these differences provide any clue about specific numbers of goslings in that year?

      (iii) Comment on anything else that you think is interesting or noteworthy about this comparison.

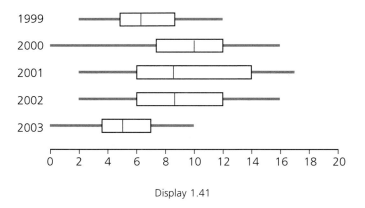

Display 1.41

7. Scientists are concerned about the layer of ozone that exists high in the atmosphere. This layer of ozone (oxygen atoms bound together in threes) filters out harmful ultraviolet light from the sun. Display 1.42 shows five-number summaries of the measurements of ozone taken during eight different years at these five stations in Antarctica: Halley Bay, Nimbus 4, Nimbus 7, Syowa, and Amundsen-Scott. Ozone is measured in units called Dobsons.

(a) Construct boxplots for each year. Use the same number line scale for all of them.

(b) What is happening to the typical ozone level during these years?

(c) What is happening to the range of measurements each year?

(d) Do you see any trend(s) in these data? If so, what? Explain briefly.

	1971	1972	1979	1980	1981	1982	1983	1984
Minimum	280	285	260	218	223	215	182	185
1st quartile	300	305	266	225	235	220	195	187
Median	305	307	280	235	245	235	205	195
3rd quartile	334	337	333	270	266	240	239	240
Maximum	370	368	367	276	312	283	245	245

Display 1.42

8. To learn the ways long trips in space might affect living things, NASA sent 13,000 tomato seeds into an orbiting platform for five years. The seeds were identical to the Rutgers tomato seeds you can buy in any hardware or department store, except that they were launched on the shuttle, orbited for five years, and then recovered. The seeds were next sent to schools all over the country to grow and to be observed. The first seedlings grown from the space seeds and seedlings from usual seeds were measured (in centimeters). Display 1.43 shows the measurements from one school.

(a) Make dotplots of the four lists of measurements. Mark the mean on the number line of each one.

(b) Make boxplots of the four lists of measurements.

(c) Are the stem lengths different between the earth seeds and the space seeds?

(d) Are the numbers of leaves different between the earth seeds and the space seeds?

(e) Suppose that the next generation of plants (grown from the seeds made by the plants just described) has the boxplots shown in Display 1.44. What can you say about these plants? Are they similar to their "parents" or have they changed? If they have changed, what do you observe about their numbers?

Three Week Old Seedlings

Earth-bound seeds		Space-exposed seeds	
Stem length	No. of leaves	Stem length	No. of leaves
29	8	26	6
20	9	29	10
23	10	22	7
23	6	26	9
26	10	28	7
27	8	17	6
22	10	25	7
18	9	21	10
30	10	30	7
28	7	16	8

Display 1.43

Three Week Old Seedlings

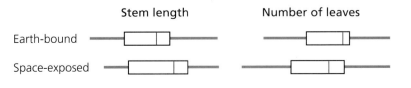

Display 1.44

1.6 Deviation

The previous section described the spread of data by looking at the range and at boxplots. It began with an example of three small data sets with the same mean, median, and range, but with different boxplots. But even boxplots can hide important differences. Data sets with similar boxplots can be spread out very differently. Here is an example:

Data Set A: 3, 3, 4, 6, 6, 6, 6, 10, 14, 15, 15, 15, 15, 16, 16
Data Set B: 3, 4, 5, 6, 9, 10, 10, 10, 10, 10, 11, 15, 15, 16, 16

We chose these numbers so that it would be easy for you to see the main idea.

1. **Find the ranges, means, five-number summaries, and boxplots of Data Sets A and B. How do they compare with each other?**

2. **Make dotplots for Data Sets A and B. How do they compare with each other?**

As you can see from your dotplots, these two data sets are spread out differently along the number line. In Set B, about half of the data is at or very near the mean. In Set A, nearly all of the data is at two values fairly far from the mean, one on each side of it. Set A is an example of a bimodal data set. Two data values that are fairly far apart have a much larger frequency than the other values. The dotplot for such a data set "peaks" in two distinct places.

There's another natural way to analyze how data are spread out. We could look at the difference between each data value and some measure of center, such as the mean. This difference is called the deviation of the value from the mean. We'll start by borrowing some "shorthand" from the graphing calculator; it makes writing these ideas a lot less tedious. Let's agree to represent the mean of a data set by $\overline{x}$. (Read "x bar.") The letter x itself will stand for any data item in the set. A letter that is used to stand for any one of a collection of numbers is called a **variable.** Now we can write the deviation of any data item, x, from the mean, $\overline{x}$, as

$$x - \overline{x}$$

How about using the average of these deviations to describe the spread of a data set?

 Try this for Data Sets A and B.

1. List and then add the deviations of these two data sets by copying Display 1.45 and filling it out. What is $\bar{x}$ in each case? Check your thinking with the few entries that have been filled in.

2. What is the average (the mean) deviation in each case? Why do you think these results worked out this way?

$\bar{x} = \underline{\ \ }$			$\bar{x} = \underline{\ \ }$	
x	$x - \bar{x}$		x	$x - \bar{x}$
3	−7		3	
3			4	
4	−6		5	−5
6			6	
6			9	
6			10	
6			10	
10			10	
14			10	
15			10	
15			11	
15			15	
15			15	
16			16	
16	6		16	6
sum:			sum:	
Data Set A			Data Set B	

Display 1.45

As you may guess from working with Display 1.45, the sum of all the deviations of a set of data is *always* zero. This is true because the mean is a balance point for the data. Not a very promising result.

But looking at differences from the mean is not a bad idea. To see how data are spread out, we can look at the distances between data points, and that's a lot like looking at the distances between individual data points and some central point. If we think of each difference as the *distance* between the data item and the mean, and if we recall that a distance is always positive or zero, then we get something worthwhile.

Display 1.46 uses Data Set A to illustrate what we are talking about.

- The first column shows each item x of data. The sum and the mean of the data are at the bottom of the column.

- The second column shows the deviations from the mean for each data item, $x - \bar{x}$. In this case, $\bar{x} = 10$, so we calculate $x - 10$. The sum and the mean of these differences are at the bottom. Both are 0.

- The third column shows what happens when we treat all these deviations as distances, which are never negative. The sum and the mean of these distances are at the bottom.

x	$x - \bar{x}$	$\lvert x - \bar{x} \rvert$
3	-7	7
3	-7	7
4	-6	6
6	-4	4
6	-4	4
6	-4	4
6	-4	4
10	0	0
14	4	4
15	5	5
15	5	5
15	5	5
15	5	5
16	6	6
16	6	6
sum: 150	0	72
mean: 10	0	4.8

Data Set A

Display 1.46

Notice the heading of the third column in Display 1.46: $|x - \overline{x}|$. The bars around the expression $x - \overline{x}$ tell us that the quantity represented is positive or zero. This is the symbol for *absolute value*.

A Phrase to Know: The **absolute value** of a number is the distance of that number from 0. It is never negative.

If a number is positive or 0, its absolute value is itself. If a number is negative, its absolute value is its opposite. Here are some examples:

$$|7| = 7 \qquad |-7| = 7 \qquad |0| = 0$$

$$\left|\frac{1}{2}\right| = \frac{1}{2} \qquad |-14.6| = 14.6$$

If absolute value signs (bars) are on the outside of an arithmetic expression, compute what's inside them first. Then make the result positive or zero. For instance:

$$|-4 + 3| = |-1| = 1 \qquad \text{but} \qquad |-4| + 3 = 4 + 3 = 7$$

Measure the length of the chalkboard in your classroom from its left end to its right end. Now measure its length from the right end to the left. Does your answer depend on the end from which you started? When finding the distance between two numbers, should it matter which one you start with?

The absolute value of the difference between two numbers is the distance between them on a number line. The order in which you take the numbers doesn't matter; the distance between them is the same.

$|12 - 3| = 9$ says that the distance between 12 and 3 is 9. (See Display 1.47.) It is the same distance as $|3 - 12|$, the distance between 3 and 12.

The Distance Between 12 and 3

Display 1.47

1. What is the distance between -5 and 2? Express this distance as an absolute value in two different ways. Draw a number line sketch of this situation.

2. What is the distance between -6 and 0? Express this distance as an absolute value in two different ways. Draw a number line sketch of this situation.

3. Explain the statement $|-6| = |6|$ in terms of distances on a number line. Draw a sketch.

Most calculators have a key or a menu choice called "abs" for finding absolute value†. The abs command uses parentheses, instead of bars, to tell the calculator that the command is to affect more than just the next number. For instance, if you enter

$$\text{abs } -4 + 3$$

the calculator will show the answer 7, but if you enter

$$\text{abs } (-4 + 3)$$

it will show the answer 1.

Thinking Tip

Estimate answers. When you use your calculator to find absolute values, try to estimate the answer in your head, too. This can help you catch mistakes in using the parentheses.

Do you remember why we started looking at absolute values? Think back about what we've been doing in this section. We want to measure how spread out a set of data is. One way to measure this is by finding the average distance between various data items and the mean. Display 1.46 shows how to calculate that average for Data Set A.

1. Copy Display 1.48, for Data Set B, which is set up just like Display 1.46. Leave room on your paper for one more column to be put in later. Then fill in the missing entries. Round your answer for the average distance to two decimal places.

2. Look at your dotplots for Data Sets A and B. In which set does the data appear to be more spread out from the center? Which set has the larger average distance? How much larger?

Keep your copy of this table for use in the next section.

This measure of spread is useful enough to have a name. It's called the **mean absolute deviation** of the data set.

† On the TI-82, it's $\boxed{\text{2nd}}$ $\boxed{\text{x}^{-1}}$. On the TI-83, it's the first choice of the MATH NUM menu.

x	$x - \bar{x}$	$|x - \bar{x}|$
3		
4		
5		
6		
9		
10		
10		
10		
10		
10		
11		
15		
15		
16		
16		
sum:	0	
mean:	0	

Data Set B

Display 1.48

Thinking Tip

Understand a complicated phrase better. Connect the meaning of each separate word with the way the phrase is defined.

Explain the meaning of each of the three words in the phrase *mean absolute deviation*. Use your explanations to make a formal definition for the whole phrase. Would you use it with measurement data? With category data?

Computer spreadsheets and some graphing calculators have a feature that makes it easy to compute things like mean absolute deviation. They allow you to store the data in a list and then automatically compute the average distances from the mean just by entering one formula. If you have such a calculator, look up the way it handles lists of data. See if you can figure out how to get it to compute the entries for Display 1.48 automatically. (For the TI-82 (TI-83), see the section of Appendix A titled Using Formulas to Make Lists.)

Problem Set: 1.6

Except for problem 1, you may use a calculator for these problems whenever it helps you.

1. Compute each of these numbers mentally, without using a calculator:

   (a) $|16 - 20|$
   (b) $|6 - 9| + |5 - 2|$
   (c) $|6 - 9| - |5 - 2|$
   (d) $|(6 - 9) + (5 - 2)|$

   Now check your results with a calculator.

2. Use your calculator to compute each of these numbers:

   (a) $|743 - 896|$
   (b) $|56 - 98| + |74 - 32|$
   (c) $|56 - 98| - |74 - 32|$
   (d) $|(56 - 98) + (74 - 32)|$

3. The following arithmetic expression doesn't tell you how to group the operations:

   $$-7 + 3 + 10 - 5 + 4 + (-8)$$

   For each part (a) – (d) place absolute value signs into the expression so that the result is the given number.

   (a) –1   (b) 27   (c) 21   (d) 13

4. The *World Almanac* lists the following monthly normal temperatures (in degrees Fahrenheit) for these three cities:

   Eureka, CA: 47, 49, 48, 49, 52, 55, 56, 57, 57, 54, 51, 48

   Grand Junction, CO: 26, 34, 42, 52, 62, 72, 79, 76, 67, 55, 40, 28

   Seattle, WA: 39, 43, 44, 49, 55, 60, 65, 64, 60, 52, 45, 41

   (a) For each set of temperatures, draw a dotplot on a number line stretching from 20° to 80°.
   (b) Find the mean monthly normal temperature for each city. Round your answers to one decimal place.
   (c) For which set of temperatures do you think the measure of spread should be largest? Smallest?
   (d) Calculate the mean absolute deviation for each set of temperatures and see if the results agree with your

answer to part (c). (Round your answers to one decimal place.)

(e) If you were selling windshield deicer, which city would you be most likely to visit on business? Which probably would not be worth the trip? Why?

5. You are a rich executive looking for a warm place to spend Thanksgiving Day. Your travel agent tells you of a remote island where the mean low temperature on Thanksgiving Day for the last ten years has been 70° Fahrenheit, with a mean absolute deviation of less than 4°. Before buying your airplane ticket, you decide to use some mathematics to see how low the temperature is likely to be during your trip.

(a) Using your calculator, verify that this is one possible set of low Thanksgiving Day temperature readings for the last ten years:

73°, 71°, 68°, 66°, 71°, 70°, 75°, 72°, 66°, 68°

(b) Show that it is possible for the island to have had a low Thanksgiving Day temperature of 55° sometime during the last ten years. (*Hint:* Experiment by changing some of the data from part (a).) Justify your answer by making up a set of 10 temperatures that includes 55° and has a mean of 70° and a mean absolute deviation of less than 4°.

(c) How low could the island's Thanksgiving Day temperature have been during the last ten years? Justify your answer by making up a set of ten temperatures that includes your answer and has a mean of 70° and a mean absolute deviation of less than 4°.

(d) Would you take the trip? Why or why not?

6. Use symbols to define *mean absolute deviation* in some efficient way. Can you find more than one way? If so, which one do you prefer, and why?

1.7 Standard Deviation

In the previous section you learned about mean absolute deviation as a way to measure the spread of data. In that phrase, the word *absolute* refers to absolute value, a way to make all differences from the mean positive.

Another way to make those numbers positive is to square them. To square a number means to multiply it by itself. It's simpler because you don't have to know in advance whether the deviation is positive or negative. The product of two positive numbers is positive, and *the product of two negative numbers is positive*, too! Of course, the size of the result will almost always be different from that of the absolute value of the number. But we just want some way to measure the spread of data, so the size change may not matter.

As you may know, the symbol for squaring is a small, raised 2 right after the number. For instance, 3^2 stands for $3 * 3$, which equals 9.

1. What is 5^2? Is it larger or smaller than $|5|$? How much larger or smaller?

2. What is $(-5)^2$? Is it larger or smaller than $|-5|$? How much larger or smaller?

3. What is $(-0.5)^2$? Is it larger or smaller than $|-0.5|$? How much larger or smaller?

4. What is $\left(\frac{1}{5}\right)^2$? Is it larger or smaller than $\left|\frac{1}{5}\right|$? How much larger or smaller?

5. $(-5)^2$ is not equal to -5^2. What's the difference? (If you think they're the same, see if your calculator agrees with you.)

6. Is $\left(\frac{1}{5}\right)^2$ equal to $\frac{1}{5^2}$? Why or why not?

Explain why these two statements are true:
- If a number has absolute value larger than 1, then its square is larger than its absolute value.

- If a number has absolute value smaller than 1, then its square is smaller than its absolute value.

Learning Outcomes

After studying this section, you will be able to:

Estimate and calculate squares and square roots of numbers

Compute the variance and the standard deviation of a set of data

Explain how variance and standard deviation describe the spread of data.

About Words

To *vary* means to change something partially or to switch around among different things. It leads to the English words *various*, *variety*, and *variegated*, as well as to the mathematical terms *variable* and *variance*.

We can measure the spread of data by taking the average (the mean) of the *squares* of all the deviations. This is an important measure of spread called the **variance**. Display 1.49 shows how to compute the variance for Data Set A from the previous section. It is a copy of Display 1.46 with one more column. That column, headed $(x - \bar{x})^2$, lists the square of each deviation. The variance is the mean of those numbers. It's at the bottom of the column.

	x	$x - \bar{x}$	$|x - \bar{x}|$	$(x - \bar{x})^2$
	3	-7	7	49
	3	-7	7	49
	4	-6	6	36
	6	-4	4	16
	6	-4	4	16
	6	-4	4	16
	6	-4	4	16
	10	0	0	0
	14	4	4	16
	15	5	5	25
	15	5	5	25
	15	5	5	25
	15	5	5	25
	16	6	6	36
	16	6	6	36
sum:	150	0	72	386
mean:	10	0	4.8	25.73

Data Set A

Display 1.49

Find your copy of Display 1.48, which you made for the previous section. Add a column headed $(x - \bar{x})^2$. Then fill in that column, including the sum and the mean at the bottom. What is the variance of this data set?

Why don't we just measure spread by taking the *sum* of the squares of the deviations? What's the point of averaging them?

In some ways, variance is easier to use than mean absolute deviation. That's because taking absolute values involves a choice based on whether something is positive or negative, but squaring doesn't. However, variance is not a good measure of the typical distance of data from the mean. (The variance in Display 1.49 is more than three times the largest deviation.) Also, data measures things in units of some kind, such as feet, hours, or degrees. Squaring gives you a measure of spread in square units. A square foot makes sense even though it's misleading here, but what's a square hour or a square degree?

The most common measure of spread fixes this problem by reversing the squaring process. This changes the variance back to the size of a typical distance from the mean and measures the spread with the original unit. The reverse of squaring is called "finding the square root."

A Phrase to Know: A **square root** of a number is a number you have to square in order to get the number you started with.

About Words

A plant grows from its *root* by watering and feeding; a number grows from its *square root* by squaring.

For instance, 9 has two square roots, 3 and −3, because $3^2 = 9$ and $(-3)^2 = 9$. In fact, every positive number has two square roots, one positive and one negative. The standard symbol for the square root is $\sqrt{}$. It's called a **radical sign**, and it's used to represent the positive square root of a number. For instance,

$$\sqrt{9} = \pm 3$$

About Symbols

The $\sqrt{}$ sign was first used in 1525. It is called a radical sign because the Latin word for root is *radix*. It was chosen to resemble a small *r*, the first letter of *radix* (and of *root*).

Sometimes the plus/minus is written in front of the radical sign as a reminder that there are two square roots, like this: $\pm\sqrt{9}$. If you ask your calculator to compute $\sqrt{9}$, it will give you 3 only. You will need to remember that −3 is also a square root of 9.

1. Find the square roots of 16. Which one is $\sqrt{16}$?

2. Find $\sqrt{5}$. Use your calculator. Is your answer exact? How do you know?

3. Estimate $\sqrt{90}$ in your head. Then use your calculator to find it. How close did you come?

4. Compute $\sqrt{\dfrac{16}{25}}$ without using a calculator.
   (*Hint:* Use what you know about multiplying fractions.)
   Check your answer using your calculator.

5. Generalize question 4 to state a rule about finding the square root of a fraction. Then try out your rule by using your calculator to find $\sqrt{\dfrac{7}{12}}$ in two different ways.

1. Compute $\sqrt{4^2}$. Then compute $\sqrt{(-4)^2}$. What's the difference between doing these by hand and using your calculator? Generalize this example to describe what happens when the calculator finds the square root of the square of a number.

2. Compute $\left(\sqrt{29}\right)^2$ without using a calculator. Generalize this example to describe what happens when you take the square of the square root of a positive number.

3. Negative numbers do not have square roots (in our usual number system). Why not?

4. Some people confuse the terms *square* and *square root*. What memory aid can you suggest to help them keep straight which is which?

We have looked at square roots because they undo what squaring does. We can use them to change the variance back to a size that's closer to a typical deviation. This gives us the most widely used measure of spread, called the *standard deviation*.

A Phrase to Know: The standard deviation of a set of data is the square root of the variance of the data.

Estimate the standard deviation of the data in Display 1.49. Is it about the size of a typical distance from the mean? Use your calculator to find the standard deviation to two decimal places.

Now let's put the ideas of this section together. Here's how to find the standard deviation of a set of data.

1. Find the mean. That is, add all the data values and divide by the number of them.

2. Find the individual deviations. That is, subtract the mean from each data value.

3. Square the individual deviations, and then add the squares.

4. Divide this sum by the number of data values. This is the variance.

5. The square root of this quotient is the standard deviation.

The 2000 Tennessee Titans had a regular season record of 13 wins and 3 losses. Their points scored were as follows:

13, 17, 23, 28, 23, 27, 14, 27,

9, 23, 24, 13, 15, 35, 24, 31

1. Follow the five-step process described above to find the mean, the variance, and the standard deviation of the Titans' scores for the 2000 season. Round your answers to one decimal place. Making a table might help you.

2. If you have a TI-82 (TI-83), enter the data into one of its lists and CALCulate the 1-Var Stats for that list. Which line of the display gives you the standard deviation? What is the calculator's symbol for it?

There actually are two slightly different ways to calculate variance and standard deviation. For reasons that you will learn when you study statistics in college, the variance is sometimes calculated by dividing by 1 *less than* the total number of data values. This affects the standard deviation because the standard deviation is the square root of the variance.

You must also know this if you use certain calculators. For instance, the TI-82 (TI-83) display *both* forms of the standard deviation. The one that comes from dividing by n (the number of data values) is called σx. The one that comes from dividing by $n - 1$ is called Sx. In this section, we prefer division by n because it better reflects the idea of averaging.

1. In the previous questions, you found the variance and the standard deviation of the Tennessee Titans' 2000 regular season point totals using division by n.

   (a) If you find the variance and the standard deviation of these scores using division by $n - 1$, will your answers be larger or smaller than before? Why?

   (b) Do it; compute these numbers to 1 decimal place using division by $n - 1$. You may use your calculators, of course. Then find the difference between these results and the previous ones.

2. Suppose you want the standard deviation for *all* the regular season scores of *all* the NFL teams in 2000. There were 31 teams and each team played 16 games.

   (a) What is *n* in this case?

   (b) Suppose the standard deviation was calculated both ways. Which of these numbers do you think would be closest to the difference between the answers: 1, 0.1, 0.01, or 0.001?

   (c) How can you use what you know about the Titans' season to check your answer for the previous question? Explain.

We end this section with a quick look at something related to what you will see in Chapter 4. To find variance and standard deviation, you begin by adding the squares of the individual deviations from the mean, right? That is, you add $(x - \overline{x})^2$ for every data value, x. Let's use the calculator's notation to write this sum as $\sum(x - \overline{x})^2$.

Now think of this sum a little differently. The mean, $\overline{x}$, is a measure of the center of the data. Suppose we replace $\overline{x}$ by some other measure of center, say the median or even some number we just guess. Can we predict whether the new sum will be greater or less than the old sum? That is, if we call this new center c, can we predict whether $\sum(x - c)^2$ will be greater or less than $\sum(x - \overline{x})^2$?

The surprising answer is Yes, *we can*. The new sum will always be greater than the old one! In other words:

The **mean** is the measure of center for which the sum of the squares of the deviations is smallest.

Sometimes this is called the "least squares" property of the mean. The following questions show you an example of this property at work.

Here are the New England Patriots' point totals for the first 13 games of the 2000 football season:

16, 19, 13, 3, 28, 24, 17, 23, 13, 11, 16, 9, 30

1. Use either the LISTs of your graphing calculator or a piece of paper to make a table with these three column headings.

$$x, \ (x - c)^2, \ (x - \overline{x})^2$$

2. List the Patriots' point totals in the x column.

3. Before you find the mean of these totals, guess a number that you think is pretty close to their center. Call that number c.

4. Now find the mean, $\bar{x}$.

5. List the squares of the deviations from c in column 2. Put their sum at the bottom.

6. List the squares of the deviations from $\bar{x}$ in column 3. Put their sum at the bottom.

7. Which of the sums of columns 2 and 3 is smaller?

8. Check with your classmates. Did any of them have a c that gave a smaller sum in column 2 than in column 3?

REFLECT

This chapter has been about data. In it, you have seen:

- ways of displaying data—bar graph, dotplot, stem-and-leaf plot, histogram, and boxplot
- ways of measuring the center—mean, median, and mode
- ways of measuring the spread—range, five-number summary, interquartile range, mean absolute deviation, and standard deviation.

Along the way, you learned about or reviewed some basic ideas about numbers, such as absolute value and square roots. These ideas appear again and again as you learn more about mathematics, skilled trades, science, or business. You also became comfortable with some parts of a graphing calculator and, through its notation, began to learn a little about algebra.

The final section of this chapter presents some projects that give you a chance to put to work many of these things you have learned.

Problem Set: 1.7

Except for problems 1 and 2, you may use a calculator whenever it helps you.

1. Do not use a calculator for this problem.

   (a) Which of the following pairs of numbers is $\sqrt{90}$ between?

   29 and 31   7 and 9   9 and 11   89 and 91

   How do you know?

   (b) Estimate the integer that is closest to $-\sqrt{59}$. Is this integer larger or smaller than $-\sqrt{59}$? Explain.

2. Compute without using a calculator

   (a) $\sqrt{376^2}$ (b) $\left(\sqrt{4.39}\right)^2$ (c) $\sqrt{(-85.01)^2}$ (d) $-\sqrt{629^2}$

3. Display 1.50 shows the diameters of the planets in our solar system.

   (a) Calculate the mean, the variance, and the standard deviation of these diameters. Name the unit of measure for each number. Round your answers to the nearest whole number.

   (b) Which unit of measure in part (a) doesn't really make sense in this setting? Explain.

   (c) Which planet has the median diameter?

   (d) Are there any typical size planets in our solar system? If so, which one(s)? Explain your answer.

Equatorial Diameters of the Planets, in Miles	
Mercury	3030
Venus	7520
Earth	7926
Mars	4217
Jupiter	88838
Saturn	74896
Uranus	31762
Neptune	30774
Pluto	1428

Display 1.50

4. The regular season point totals for the 2000 San Francisco 49ers were

   28, 22, 24, 41, 27, 28, 28, 16, 24, 15, 21, 16, 45, 27, 17, 9

(a) Find the mean and the standard deviation of the 49ers' scores for the 2000 season. Round your answers to one decimal place.

(b) Compare your answers with the mean and standard deviation for the Tennessee Titans' 2000 season, which you found earlier. Which team had the higher mean score? Which team had the larger standard deviation?

(c) Which offense was more consistent in its scoring? Explain.

5. Display 1.51 lists the all-time top 20 money making American movies, as of January 13, 2005. What is the mean dollar amount for these 20 movies? How many of these movies differ from the mean dollar amount by less than one standard deviation (larger or smaller)?

Rank	Movie (date)	Amount
1.	Titanic (1997)	600.8
2.	Star Wars (1977)	460.9
3.	Shrek 2 (2004)	436.5
4.	E.T. : The Extra Terrestrial (1982)	434.9
5.	Star Wars Episode I: The Phantom Menace (1999)	431.1
6.	Spider-Man (2002)	403.7
7.	LOTR: The Return of the King (2003)	377.0
8.	Spider-Man 2 (2004)	373.4
9.	The Passion of the Christ (2004)	370.3
10.	Jurassic Park (1993)	357.1
11.	The Lord of the Rings: The Two Towers (2002)	340.5
12.	Finding Nemo (2003)	339.7
13.	Forrest Gump (1994)	329.7
14.	Harry Potter and the Sorcerer's Stone (2001)	317.6
15.	The Lord of the Rings: The Fellowship of the Ring (2001)	313.8
16.	The Lion King (1994)	312.9
17.	Star Wars Episode II: Attack of the Clones (2002)	310.7
18.	Return of the Jedi (1983)	309.1
19.	Independence Day (1996)	306.1
20.	Pirates of the Caribbean (2003)	305.4

All-Time Top Money Making American Movies[4]
(Amounts are in millions of dollars.)

Display 1.51

[4] From _Top Grossing Films: U. S.,_ by The Movie Times. See www.the-movie-times.com.

1.8 Some Projects

Learning Outcomes

After studying this section, you will be able to:

Design and carry out a project that requires data collection

Construct data displays and calculate summary statistics

Interpret and describe the meaning of the data

Justify conclusions drawn from the data.

I—HEARTBEATS

Everyone probably knows that your heart beats 72 times per minute. People believe this even if they haven't taken their pulse lately (or maybe ever). Can you trust numbers like these? Where do people get them? Here's a chance to find out what's really going on.

- Have everyone in the class measure and record their pulse rates – how many times their heart beats in a minute. You can feel your pulse at the inside of your wrist, or just under the corner of your jaw (use your fingertips, not your thumb for these), or sometimes with your hand over your heart. Count the number of beats in a full minute. Pool your data with that of the rest of the class.

- Create these from the data: a dotplot, a stem-and-leaf plot, a histogram, and a boxplot.

- Calculate the mean, median, range, five-number summary, interquartile range, and standard deviation.

- Use the results of your summaries and calculations to help you answer these questions:

  – What is the typical pulse rate for your class? Which of the measures of center do you think is the best? Why?

  – How spread out are the data? Which number or calculation do you think gives the best estimate of the spread? Why?

- Next, measure the pulse rate of another person in your home. Bring the data to class and make a separate data pool.

- Calculate the mean, the median, and the standard deviation, and construct a boxplot for this new set of data.

- Compare the two sets of data (from class and from home). What can you say about the centers? What can you say about the spread? If there are differences, make a list of the things that might explain them. (For instance, pulse can be affected by height, weight, activity, physical fitness, diet, etc.) Write a paragraph that summarizes the results of your comparison.

II—READABILITY

One way of deciding the readability level of books, magazines, newspapers, etc., is by looking at the number of words per sentence. The fewer words per sentence, the lower the readability level. Other things (such as word length) are counted, too, but sentence length is an important factor. In this project we assume that readability level depends mainly on sentence length.

Here are some questions about readability level that you can look at by using the data analysis tools of this chapter.

1. How does the readability level of your local newspaper compare with that of *USA Today*?

2. How does the readability level of your school newspaper compare with that of *USA Today*?

3. How does the readability level of your school newspaper compare with that of your local newspaper?

4. Is there a difference between the readability level of the news section and that of the sports section (in whichever newspaper you choose)? Between the news and the business sections? Between the sports and the business sections?

5. Is the readability level of your (school or local) newspaper consistent from issue to issue? Is there a change from one month to the next? Or, particularly in the case of your school newspaper, does it change when a new editor or a new faculty advisor is appointed?

To deal with any of these questions, you first have to decide how to interpret words per sentence.

- Should you use the mean or the median? Maybe you ought to look at both.

- Will dotplots, histograms, or stem-and-leaf plots give you any useful information? If so, what?

- Will boxplots give you any useful information? If so, what?

- Will standard deviation give you any useful information? If so, what?

Next, there's the question of gathering the data.

- Are you going to count all the words in every sentence? That's a long, tedious job!

- If you don't count words in all the sentences, which sentences will you pick?

- If you just pick out some sentences, will the way that you have decided to pick them give you a fair sample of the entire newspaper (or section)? Why or why not?

Here's one approach to some of these questions:

1. Pick out two newspapers (or two sections of a newspaper) that you want to compare.

2. Decide on a way of picking out a sample of 40 sentences from each.

3. Count the words in each sentence you pick, then compute the five-number summary for each 40 sentence sample.

4. Make a boxplot for each data set, using the same scale.

5. Discuss whether or not your results indicate that the two newspapers (or sections) have different readability levels.

6. Compute the mean and the standard deviation for each 40 sentence sample and see if the results support your conclusions.

III—PLANTS

Here's a chance to generate some data and watch them change, as well.

- Get a flat of soil and a few packets of cherry tomato or plum tomato seeds. Sow 50 seeds and keep them watered and tended.

- As soon as you get seedlings, measure their heights (in centimeters) twice a week for three weeks. You should end up with six sets of measurements. Each time you measure, take care to note which seedlings are unusually short or unusually tall.

- Enter your six measurement sets into the data storage lists in your graphing calculator. Use the calculator to construct a boxplot for each data set. Plot them on the same screen and compare them.

1. Now write a paragraph or two answering these questions:

   (a) What happens to the median heights as the weeks go by?

   (b) What happens to the spread of the data?

   (c) Were there any very short or very tall seedlings in the first week?

   (d) Were there any very short or very tall seedlings in the last week? Were they the same as the very short or very tall seedlings in the first week?

2. Raise the plants until they bear fruit. Keep them in full sun and make sure they have large enough pots. Then harvest the fruit. Yes, tomatoes are fruit. Construct a boxplot of the weights of the tomatoes you harvest. Describe the measures of center and spread of the data; then be sure to eat the fruits of your labors.

IV—A MYSTERY
The goal of this project is to solve a small mystery.

- When an old book or manuscript is discovered, sometimes the identity of the author is not certain. The people who study such things can often narrow the possibilities to two or three possible authors.† In these cases, one way of getting a clue to the identity of the real author is to use word frequency patterns.

- The idea is that when we write, each of us tends to use the most common words—*and*, *the*, *a*, etc.—with frequencies that are predictable in the long run. This pattern of word use is a sort of fingerprint of a person's writing style. So, one way to help decide whether a particular person wrote a particular book is to compare the word use patterns of the person's known writing with the word use patterns in the book. Computers have made this task much easier in recent years.

---

†This kind of question has arisen about things that may have been written by women during times when women's writing was not valued. To give the work a better chance of being published, a male author might have allowed his name to be used.

• This project lets you try your skill at fingerprinting someone's writing. Your teacher will divide your class into groups and give each group the names of two authors whose books are in your school library. You will also be given some pages copied from another book which is by one of these authors, but is not in your school library. Your job is to try to decide from a word pattern which author wrote these mystery pages, as follows:

1. Choose one book by each author.

2. For each book, choose twenty-five 60-word samples of the writing, count how often the word *and* occurs in each sample, and make a list of these numbers. To choose a sample, just turn to a page at random, pick a place on it, and start counting words from there until you get to 60. These two 25 number lists are your data sets for the two authors.

3. Find the mean, the standard deviation, and the five-number summary for each list. A spreadsheet program or a graphing calculator can help you do this part quickly and easily.

4. Display the data of each list as a histogram and as a boxplot.

5. Now decide how to choose samples of how often *and* occurs in the mystery pages. Once you have decided, make a list of your data, find its mean, standard deviation, and five-number summary. Then display the data as a histogram and as a boxplot.

6. Compare your information about the mystery pages with your data about the two authors; then decide which one is more likely to have written the mystery pages. Write a paragraph justifying your decision.

5/07/98 reading (Estimated) 819
4/08/98 reading (Estimated) 816
Meter reading difference   22
Meter multiplier   ×10
Total KWH used in 29 days   220

CHARGES FOR ELECTRICITY USED

Basic service charge: $7.18
  (does not include usage)

           KWH   COST/KWH
First   220.0 @ 13.6909¢   30.12
   Adjustment @   .6909¢   1.52
Tax @ 4.000%   1.55

George Cooper
Helping to Keep Down Electricity Costs

George Cooper is a Customer Service Manager for Entergy, a world wide energy company based in the southern U.S. George works with whole cities and towns to guide them in the wise use of energy. He also helps customers with high bill problems.

George went to Alcorn State University in Mississippi where he studied business administration. At Entergy, before becoming a Customer Service Manager, he was an auditor. "I had to evaluate each department," he explains. "This included areas like Customer Accounting, Meter Reading, and Engineering. In Customer Accounting, for example, I would pull customer accounts and calculate their bills manually. This way, I'd check on computer accuracy."

A number of variables go into the equations that determine a customer's bill. "There is the basic customer charge, whether the electric service is overhead or underground and different rate schedules. For example, the first 500 kiloWatt-hours are based on higher rates than all additional kiloWatt-hours. And certain billing months cost more than others. Even the fuel adjustment costs vary monthly. We program these equations into computers. Then we check them regularly for error."

When a customer has a high bill complaint, George looks for the root cause. One customer, he recalls, "couldn't figure out why their bill had tripled in just a thirty-day period. We checked the meter and it was accurate. So I met with the customer's electrician and we found where duct work in the attic had come loose. That's where they were losing the heat. Our policy is to work with customers until we find the answer."

Welcome to Algebra

2.1 Abbreviations All Around Us

Learning Outcomes

After studying this section, you will be able to:

Describe the advantages and disadvantages of abbreviations

Use algebra to express relationships between sets of numbers

Use relationships abbreviated by formulas to determine unknown values.

Long ago, when one of the authors of this book was a little boy, he asked his father what algebra was. His father said that algebra was about figuring out what x is. The little boy thought this was very strange. Why are millions of young people each year trying to figure out what x is? Why not ask some genius to settle the question once and for all?

You can tell that the father in this story did not give his little boy a clear idea of what *algebra* is all about. But you will learn more about algebra as you read this chapter.

The most obvious difference between algebra and arithmetic is that arithmetic uses specific numbers, such as 34, 9.3, or $\frac{1}{2}$, while algebra also includes letters, such as x or y or z. These letters are used as abbreviations. Let's look at some other familiar abbreviations.

When the Post Office introduced ZIP codes in 1963, it also adopted two-letter abbreviations for each of the fifty states. For example, CA is the abbreviation for California and FL is the abbreviation for Florida.

1. What are the advantages of the two-letter abbreviations?

2. What are the disadvantages of the two-letter abbreviations?

3. How many of the two-letter state abbreviations do you know? Write them down and compare your list with the lists of your classmates. Did your class get all 50 states?

One common type of abbreviation is the **acronym.** Acronyms abbreviate a string of words usually using the first letter of each word. For example, U.S.A. stands for <u>U</u>nited <u>S</u>tates of <u>A</u>merica. Often, short words like "of" are left out of the abbreviation. Another example of an acronym is SD for South Dakota.

a

Which of the two-letter state abbreviations you and your classmates found are acronyms?

The Post Office also approved some additional two-letter abbreviations that are acronyms. For example, PR stands for Puerto Rico and D. C. stands for the District of Columbia.

b

Here are some other acronyms. For each one, find the words that the letters stand for.

NAACP
NCTM
ASAP
RSVP

Compare your answers with those of your classmates. Discard any answers you agree are wrong. Were there any acronyms that stood for more than one thing?

Another kind of abbreviation uses parts of a phrase. The programming language Fortran is an abbreviation for <u>for</u>mula <u>tran</u>slation. Fortran translates engineering formulas into a form the computer can understand.

Find out what the following abbreviations stand for. Are any of them acronyms?

NASA
C. I. D.
IOU

One advantage of abbreviations is that they take less time to write down!

Four friends stop at the local diner for lunch. The waiter arrives to take their order. Ani wants two cheeseburgers, a large order of fries, and a medium cola. Carlos orders a bacon double cheeseburger, a medium order of fries, and a large orange drink. Juanita wants a fish sandwich, a large order of fries, and a large diet cola. Soo orders a chicken sandwich, a large fries, and a medium root beer.

1. If you were the waiter, what would you write down on your order pad? Remember that you want to write as little as possible and yet be clear to the person in the kitchen who must fill the order.

2. Assign prices to the various items ordered and write up a check for the order.

3. Finally, write an expression for the total cost of the order using the shorthand that you created for the various food items. Again, remember that you want it to be brief and clear!

Acronyms and ordinary abbreviations are made up of letters, symbols, and/or numerals. For example, H_2O. Mathematics has its own abbreviations and symbols. We can think of 23 as an abbreviation for twenty-three. We also have very short symbols for operations such as $+$ and $-$. There are symbols for relations between numbers, such as $=$ and $<$. You have seen all these symbols in your study of arithmetic. Algebra adds letters to the list of symbols. Unlike acronyms, which abbreviate words and phrases, letters in algebra represent numbers. One use of these abbreviations is to represent relationships we find in the world around us. These relationships can be described in words, but we can describe them much more briefly with algebra.

0	4
1	5
2	6
3	7
4	8

Display 2.1

Describe the patterns in Display 2.1. How are the numbers in the second column related to those in the first?

We can use algebra to express the pattern in Display 2.1. First, we label the columns with letters.

x	y
0	4
1	5
2	6
3	7
4	8

Display 2.2

Here we use the letter x as an abbreviation for "a number in the first column" and the letter y as an abbreviation for "a number in the second column." We could have used other letters instead of x and y (for example, r and s). Particular numbers that show up in the table are called **values** of the variables x and y. Recall, a letter that is used to stand for any one of a collection of numbers is called a *variable*. The values of y in our table are 4, 5, 6, 7, and 8. When we have more than one variable, it is common to want to study the relationships between the two. Now we can abbreviate the relationship between x and y in Display 2.2 as

$$y = x + 4$$

Use algebra to describe the relationship between u and v in Display 2.3.

u	v
0	0
1	2
2	4
3	6
4	8

Display 2.3

The pattern in Display 2.3 could be expressed using multiplication or division. If we use multiplication, we need to be careful how we write it. One common symbol for multiplication is ×. We could write $v = 2 \times u$. A problem with this is that the × could be a letter or it could be a multiplication sign. These are hard to tell apart when we write them by hand or on a typewriter. Because × is often used to name a variable, we usually use some other symbol for multiplication in algebra. In spreadsheets and computer programming languages, the usual symbol is *. We could write our pattern as

$$v = 2 * u$$

Once we have found a pattern in a table, we can use it to find out things that were not in the original table. Because the abbreviations of algebra represent numbers, we can calculate with them. For example, in the first table, Display 2.2, we had $y = x + 4$. Although the value 53 for x did not appear in the table, we can use the pattern to calculate the y that goes with $x = 53$ if the pattern continues.

$$y = x + 4 = 53 + 4 = 57$$

Find the value of v when $u = 23$ in Display 2.3.

a

Formulas are abbreviations that use algebra. For example, the area of a rectangle is given by the formula

$$A = l * w$$

A is an abbreviation for the area, l is an abbreviation for the length, and w is an abbreviation for the width.

Find the area of a rectangle 4 meters by 8 meters.
What is the unit of measure in your answer?

b

We can use algebra to abbreviate laws. For example, you probably know the **Commutative Law of Addition.**

$$a + b = b + a$$

What is special about a law that it is true for *all* values of its variables?

Can you find any values of *a* and *b* that make

$$a + b = b + a$$

a false statement? Can you find any values of *a* and *b* that make

$$b = a + 4$$

a false statement? Is this a law? Explain.

The rest of this chapter will show you many more examples of using algebra to abbreviate patterns, relationships, laws, and formulas.

Problem Set: 2.1

1. Use algebra to describe the relationship between the variables in each table.

x	y	a	b	c	d	k	z
0	0	0	13	5	32	5	65
1	7	1	14	6	33	6	78
2	14	2	15	7	34	7	91
3	21	3	16	8	35	8	104
4	28	4	17	9	36	9	117

Display 2.4　　　Display 2.5　　　Display 2.6　　　Display 2.7

2. Suppose you go into a store to buy some loaves of bread. The bread costs $1.39 per loaf. If you buy five loaves, how much will they cost? If you buy eight loaves? How about ten loaves? Make a table summarizing your results. Now write a formula for the cost C in dollars of the number *n* of loaves.

3. If you travel at 40 miles per hour, how far do you travel in one hour? In two hours? In five hours? Now write a formula for distance (*d*) in a given time (*t* hours) when you travel at a constant rate (*r*) of miles per hour. Use your formula to find out how far you travel in nine and one-half hours.

4. The absolute temperature scale is used in scientific work. It uses a degree the same size as the Celsius degree, but starts at absolute zero rather than the freezing point of water. Temperatures on this scale are said to be in degrees Kelvin. To find the absolute temperature A in degrees Kelvin, just add (approximately) 273 to the Celsius temperature T:

$$A = T + 273$$

(a) Normal room temperature is about 20 degrees Celsius. What is the corresponding absolute temperature?

(b) Water boils at 100 degrees Celsius. What is the corresponding absolute temperature?

(c) If the temperature is 0 degrees Kelvin, what is the Celsius temperature? What sort of weather would this be?

(d) If the temperature is 300 degrees Kelvin, what is the Celsius temperature? What sort of weather would this be?

5. The perimeter P of a square s units on a side is given by the formula

$$P = 4 * s$$

(a) What does *perimeter* mean?

(b) Find the perimeter of a square 15 feet on a side.

(c) Find the perimeter of a square 7 meters on a side.

(d) A square has a perimeter of 32 meters. How long are the sides? How do you know?

6. Mr. Sen has a workshop in his garage. There is no electricity in the garage, so he runs a 100 foot extension cord from his house to the workshop when he wants to use electrically powered tools. He has noticed that his electric drill works fine but his saw does not. An electrician friend suggested he apply **Ohm's Law**. This relates the voltage drop or loss, V, to the current flowing through the extension cord and the resistance, R, of the cord.

$$V = i * R$$

His extension cord has a resistance of $R = 1$ ohm. If he knows the current, i, through the cord, he can calculate the voltage drop, V, along the extension cord. For example, a 60-watt light bulb uses about 0.5 amperes of current. Then

$$V = i * R = 0.5 * 1 = 0.5 \text{ volts}$$

is the voltage drop due to the extension cord. If the voltage in the house is 117 volts, the voltage at the end of his extension cord will be $117 - 0.5 = 116.5$, which will still be enough to light the light bulb.

(a) What will the voltage drop be for a drill that uses 3 amperes?

(b) What will the voltage be at the end of the extension cord?

(c) The drill is rated to work with voltages of 110–120 volts. Would you expect the drill to work well at the end of this extension cord?

(d) What will the voltage drop be for a table saw that uses 10 amperes?

(e) What will the voltage be at the end of the extension cord?

(f) The saw is also rated to work with voltages of 110-120 volts. Would you expect the saw to work well at the end of this extension cord?

(g) What do you think Mr. Sen should do if he wants to use the table saw?

2.2 Algebra Is Abbreviations

Here is a simplified version of a bill for one month's use of electricity by a household.

CENTRAL CONNECTICUT ELECTRIC COMPANY

Charge to:　　　　John P. Smith,
　　　　　　　　　54 Cromwell Place, Meriden, Connecticut 06496

For Services At:　54 Cromwell Place, Meriden, Connecticut 06496

Billing Period		Meter Reading		Number of KiloWatt-Hours Used
From	To	Previous	Current	
April 1	April 30	7553	7928	375

This is your electric bill calculation:

Customer Service Charge	$ 8.50
Energy Charge 375 kWh x $0.09	$33.75
Total Charge (amount now due)	$42.25

Display 2.8

Before discussing the electric bill, we want to be sure that you understand some basic terms.

Use a dictionary, or ask a science teacher or your mathematics teacher the meanings of *watt, kiloWatt,* and *kiloWatt-hour.* Discuss these terms with your classmates to make sure you understand what these terms mean.

Write a short paragraph explaining in your own words the meanings of *watt, kiloWatt,* and *kiloWatt-hour.*

Now we know that a kiloWatt-hour is a measure of quantity of electricity, or of electrical energy. (A kiloWatt is a measure of power, the rate of usage of electricity.) Just as time * rate = distance,

$$t * r = d$$

time (t) times the rate at which electrical energy is used (P) equals total energy use (E).

$$t * P = E$$

Learning Outcomes

After studying this section, you will be able to:

Distinguish between a variable and a constant

Use variables and algebraic equations to describe real world relationships.

About Symbols

kWh is an abbreviation for kiloWatt-hour or kiloWatt-hours.

How much electrical energy is used by a
- 150-watt light bulb turned on for 12 hours?

- 2750-watt electric space heater turned on for 3 hours?

- 15-watt night light burning 8 hours per night for 30 nights?

- 1250-watt hairdryer used 10 minutes per day, every day for a year?

Can you explain the computation of the electric company's bill for $42.25, as noted in Display 2.8? If so, write your explanation. If not, explain where you are having trouble.

There is a charge of 9 cents per kiloWatt-hour used, so this part of the bill is figured as

$$375 \text{ (kWh)} * 0.09 \left(\tfrac{\text{dollars}}{\text{kWh}}\right) = 33.75 \text{ (dollars)}$$

In addition to paying for the amount of electricity used, the customer pays a service charge of $8.50 per month. This is a fixed charge that the customer has to pay even if no electricity is used during the month. Thus, the total charge equals

$$33.75 \text{ (dollars)} + 8.50 \text{ (dollars)} = 42.25 \text{ (dollars)}$$

Note that to compute the bill of $42.25, we needed nothing but arithmetic: multiplication and addition of known numbers.

			METER READING			
Customer Name	Previous (March)	Current (April)	# kWh Used	Charge per kWh	Cst Svc. Charge	Total Charge
Smith	7553	7928	375	$0.09	$8.50	____
Abel	3214	3624	____	$0.09	$8.50	____
Hsu	8327	____	225	$0.09	$8.50	____
Charles	____	10,516	280	$0.09	$8.50	____
Dahl	____	____	416	$0.09	$8.50	____
Estephen	9824	____	____	$0.09	$8.50	$34.60
Fox	4528	4708	____	$0.09	$8.50	____
Georgis	5265	5505	____	$0.09	$8.50	____

Display 2.9

But now let us consider bills for the same one month period for eight customers. We will simplify the bills further by setting out each bill under seven headings. Your teacher will give you a copy of Display 2.9. Fill in the blanks.

a

As was the case for John Smith's bill, we need nothing but arithmetic to fill in the blanks in Display 2.9. What we would like to do now is to obtain an expression to summarize how the bill is computed for *any* number of kWh used, not just for the Smith household usage of 375 kWh.

Can you come up with such an expression? If so, do it. If not, explain where you're having difficulty.

b

An expression of this type is most easily obtained with the help of abbreviations. One way to do it in this case is to abbreviate the number of kWh used (the usage), by a single letter, say u. Then, because there is a charge of 9 cents (or 0.09 dollars) per kWh used, the energy charge (in dollars) equals $0.09 * u$. Hence, the total charge equals $0.09 * u + 8.50$. If we let T stand for the total charge (the amount due), we obtain the equation

$$T = 0.09 * u + 8.50 \text{ (dollars)}$$

A Word to Know: An **equation** is a symbolic statement that two quantities are equal.

> To summarize the utility example,
> If u = the number of kWh used, and
> T = the total charge in dollars,
> then $T = 0.09 * u + 8.50$ (dollars).

There are some important things you should notice about this summary.

(1) The letters in an algebraic equation stand for *numbers*, so avoid "let u be kWh". Be careful to write "let u be the number of kWh used".

(2) In the equation expressing T, we have written "(dollars)" after the equation to remind us of the *units* that we chose to use for the total charge. We could, of course, choose other units, such as cents.

2.2 Algebra Is Abbreviations

a

If *C* equals the total charge in cents, write an equation expressing *C* in terms of *u*.

A big advantage of the equation, $T = 0.09 * u + 8.50$ (dollars), is that we can use it to compute the electric bill for any number of kWh used, not just for the electric usages in Display 2.9.

We call *u* a *variable* because the usage varies from one household to another. Likewise, *T* is a variable.

In contrast, 8.50 and 0.09 are called **constants** — they remain fixed throughout the discussion.

Looking at Display 2.9, how can you tell the constants from the variables?

About Symbols

Ellipsis points, "...", are used to indicate something is missing. For example, we might say, "The natural numbers are 1, 2, 3, ..." In Display 2.10, we have used three vertical dots to indicate that many other people received electric bills that month.

We summarize the essential features of Display 2.9 in Display 2.10.

u (kWh)	*T* (dollars)
375	42.25
410	45.40
225	
280	33.70
416	
290	34.60
	24.70
	30.10
⋮	⋮
u	$0.09 * u + 8.50$

Display 2.10

The expression at the bottom of the table, $0.09 * u + 8.50$, abbreviates the pattern that all the bills follow.

b

Find the missing entries in Display 2.10. Use the version of Display 2.9 you have already filled in.

Now the electric company has requested permission of the Public Utility Commission (PUC) to boost its usage rate from 9 cents per kWh to 10 cents per kWh. After public hearings on the subject, the PUC has authorized this increase, provided that the customer service charge is reduced from $8.50 to $6.00. The company accepts this compromise.

Letting N stand for the total charge in dollars under this new rate schedule, write an equation expressing N in terms of the usage, u (kWh). Your teacher will give you a copy of Display 2.11. Fill in the total charge for each usage under the new rate schedule.

Usage, u (kWh)	Old Rate Schedule Total charge T (dollars)	New Rate Schedule Total charge C (dollars)
375	42.25	
410	45.40	
225	28.75	
280	33.70	
416	45.94	
290	34.60	
180	24.70	
240	30.10	
•		
•		
•		
u	$0.09 * u + 8.50$	$0.10 * u + 6.00$

Display 2.11

When you finish your table, you should find that sometimes the bill under the new rate schedule is greater than under the old rate schedule; in other cases, the new one is less than under the old one.

1. Describe for which values of u the old rate schedule leads to a smaller bill than the new rate schedule and for which values of u the new rate schedule leads to a smaller bill than the old rate schedule.

2. Do you think that one of the rate schedules is less expensive than the other? Explain your answer.

3. Find a particular usage, u (kWh), for which old and new rate schedules lead to exactly the same total

charge. If you succeed, state the value of u that leads to the same total charge, what that total charge is, and how you found this usage and total charge. If you did not succeed, state in words what you attempted.

4. Looking at Mr. Smith's usage of electricity for the eight months in Display 2.11, do you think that any of his bills under the new rate schedule would be less than under the old one? Explain your reasoning.

In this section you have seen two examples of using algebra to abbreviate a pattern in the world around us. Usually we start with a situation in the real world that we want to analyze, like the electric bills. We identify some variables of interest, and then choose letters to represent these variables. Although we can use any letter to represent a variable, some choices are more helpful than others. If possible, choose a letter that will help you remember what variable it represents. For example, we used T for total charge, in dollars; u for electric usage, in kWh; Q for quantity of oil, in gallons; and so forth. State clearly what each letter you have chosen represents. Remember that the letter must represent a number. In a problem about buying loaves of bread, you could let C represent the cost in dollars of your purchase of some loaves, n represent the number of loaves bought, and p represent the price of one loaf. Remember that variables represent *numbers*. Don't write that a variable stands for "loaves," because loaves are not numbers! Whenever measurements are involved, be sure to state the units you have chosen.

Problem Set: 2.2

1. A German shepherd puppy weighs about 1 pound at birth and gains about 5 pounds per month during the first year.

   (a) If W stands for the weight in pounds (lb.) of a puppy aged m months, copy and complete the following table.

m (months)	0	1	2	3	4	5	6
W (lb.)							

Display 2.12

(b) Write an equation for the weight W (lb.) of a puppy of age m (months).

(c) In your answer to part (b), which are the variables and which are the constants?

2. Saunders and Daughters Fuel Company delivers home heating oil for $1.149 per gallon, plus a charge of $5 for the delivery.

   (a) If P stands for the price in dollars for the delivery of g gallons of oil, copy and complete the following table:

g (gallons)	100	200	300	400
P (dollars)				

Display 2.13

   (b) Write an equation for the price of a delivery of any quantity of oil.

   (c) In part (b), which are the variables and which are the constants?

   (d) Suppose that we change the units of price from P (dollars) to C (cents). Write the equation for C in terms of P.

   (e) Suppose that we change the unit of quantity of oil from g (gallons) to q (quarts). Write an equation for the price P (dollars) in terms of the quantity q (quarts) of oil delivered.

3. Look for a pattern in the equations of our electric bill example, the puppy weights in problem 1, and the oil prices in problem 2. Describe the pattern in sentences, including mathematical symbols, if you wish.

4. (a) Find the minimum, median, mean, maximum, and range of the data set consisting of the eight values of u in your copy of Display 2.10.

   (b) Find the minimum, median, mean, maximum, and range of the data set consisting of the eight values of T in your copy of Display 2.10.

   (c) Write an equation for each answer in (b) that shows how it is related to your answer in (a).

5. Suppose that Mr. Smith's usage of electricity for each month of the year varied, as in Display 2.14.

Month	Usage u (kWh)
January	430
February	395
March	395
April	375
May	350
June	340
July	510
August	515
September	380
October	375
November	395
December	425

Display 2.14

(a) Suggest some reasons for the variation in electrical usage from month to month.

(b) Make a bar graph to exhibit the electrical usage throughout the year. The electric company usually prints such bar graphs on customers' monthly bills.

(c) Copy Display 2.14 and add a third column with the heading "Total Charge T (dollars)". Fill in this column by calculating the total charges, using $T = 0.09 * u + 8.50$.

(d) For the data set consisting of the 12 values of T calculated in (c), calculate mean and median T.

(e) Consumers sometimes avoid the strain of having large electric bills at certain times of the year by arranging with their electric company to pay the same amount for each of the 12 months of the year. To do this, the company assumes that the bills for the previous year are typical of the usage for the current year. What is Mr. Smith's total cost for the entire year? What must he pay each month if he wants equal payments with the same total?

6. Display 2.15 is a picture of a long rod used in surveying. There is a mark on the rod x units from one end and y units from the other end.

Display 2.15

(a) If $x = 1.2$ m, write an expression for the length of the rod.

(b) Write an expression for the length of the rod if you don't know the value of x.

(c) If the total length of the rod is 3 m, write an equation representing what you know from the diagram.

7. The air temperature, T (in degrees Celsius) at an altitude A (meters above ground level), is usually estimated by dividing A by 100 and subtracting the result from the ground temperature, G (in degrees Celsius). Write an equation showing how the air temperature drops as you rise in altitude.

8. For the displays in (a) and (b) below,

- describe in words how the variable in the second column relates to the variable in the first column;

- choose appropriate letters to represent the variables, as needed; and

- write an equation that summarizes the relationship, if you can. If you cannot, then explain why not.

(a) A bacterial culture is being treated with ultraviolet rays in an attempt to kill the bacteria (Display 2.16).

Use your equation to estimate what percentage would survive after four hours. Do you think your equation could be used to predict what percentage would survive after 24 hours? If so, do it. If not, explain why not.

Treatment Length (in hours)	Percent Surviving
0	100
1	91
2	82
3	73

Display 2.16

(b) A typesetter needs to know how many spaces are needed to print various numerals (Display 2.17).

Numeral	Number of Spaces
12	2
−12	3
1.2	3
$\frac{1}{2}$	3
0.0000034	9
3400000	8

Display 2.17

9. Here is some practice in writing algebraic equations as abbreviations for sentences. Choose appropriate letters to represent the variables in each problem. Be sure to write down what each letter represents.

(a) Write an equation for the cost of a telephone call if the charge is 78¢ for the first minute plus 69¢ for each additional minute.

(b) The perimeter of a rectangle can be found by adding its length, its width, its length, and its width.

(c) John returns aluminum cans for 5¢ each and large glass bottles for 10¢ each. Write an equation that expresses the total cash return on a collection of cans and bottles.

(d) If Lisa is paid $40 per week plus 1.5% of the value of the merchandise she sells, write an equation for her weekly earnings.

(e) Write an equation for the speeding fine in Illinois, where the speed limit is 55 mph and the fine is $75 plus $5 for each mile per hour over the speed limit.

(f) Write an equation for systolic blood pressure using the estimation rule: age divided by 2, plus 100.

(g) Write an equation for the maximum exercise pulse rate using the estimation rule: age in years multiplied by 0.8, with the result subtracted from 172.

(h) The total amount of money that McBurger's takes in on its sales is the number of chili burgers sold multiplied by $2.50, the price of a single chili burger. Write an equation for the total amount of money McBurger's takes in for chili burgers.

(i) We also know that the amount of sales tax that McBurger's has to pay on its sales is found by multiplying its sales by the state sales tax rate. Write an equation for the sales tax paid on a day's sales of chili burgers.

10. Instead of buying a car, you wish to lease one for a specified number of months. You have your heart set on a red Wombat convertible.

(a) One dealer, Jolly Jack, offers to lease you one of these cars for an initial (one time) charge of $3,000, plus a monthly charge of $299, for 24 months. You must agree to pay for the whole two-year period.

    (i) Write an equation for the amount that you will have paid Jolly Jack at the end of any number of months within the lease period.

    (ii) How much will you have paid by the end of the first year? By the end of the lease?

(b) A rival dealer, Smilin' Sam, offers to lease you an identical car for an initial charge of $2,400, plus a monthly charge of $340, for 24 months.

    (i) Write an equation for the amount that you will have paid Smilin' Sam at the end of any number of months within the lease period.

    (ii) How much will you have paid by the end of the first year? By the end of the lease?

    (iii) Do you prefer Jolly Jack's offer or Smilin' Sam's? Explain why.

(c) At Happy Helen's, you pay the same total amount as at Jolly Jack's for only $2,200 down for leasing an identical car for 24 months.

    (i) What is Happy Helen's monthly charge?

    (ii) Do you prefer Jolly Jack's offer or Happy Helen's? Explain why.

PROJECT

Look at last month's electric bill for wherever you live. What was the total usage? What would that cost under the rate schedules we have studied in this section? See if you can figure out the rate schedule for your own bill. How does it compare to the rates in this chapter? If you can, get bills for the past year. How does the usage vary with the seasons? Do you think the pattern would be the same in Alabama and Alaska? Explain.

2.2 Algebra Is Abbreviations

2.3 Laws of Algebra

Learning Outcomes

After studying this section, you will be able to:

Explain the Commutative and Associative Laws of Addition and Multiplication

Express the Commutative and Associative Laws of Addition and Multiplication in algebraic symbols

Explain the use of parentheses to indicate the order in which operations are performed

Apply the conventions on the order of operations to evaluate algebraic expressions

Explain the Distributive Law relating addition and multiplication

Express the Distributive Law in algebraic symbols.

In the previous section, the equation $T = 0.09 * u + 8.50$ expressed a relationship between the number of kiloWatt-hours used, abbreviated by u, and the total charge for electricity, abbreviated by T. This is an example of using algebra to describe relationships in the world around us. Another use of algebra is in expressing mathematical facts.

Here we have two *laws* of arithmetic expressed in algebraic notation.

Commutative Law of Addition $\qquad a + b = b + a$
Commutative Law of Multiplication $\quad a * b = b * a$

A **law** is an equation that is true for *all* values of its variables. The general law $a + b = b + a$ is an abbreviation for many specific facts, such as

$$2 + 3 = 3 + 2$$
$$2.7 + 3.12 = 3.12 + 2.7$$
$$1234 + 4321 = 4321 + 1234$$

Determine whether or not the pairs of activities listed are commutative, that is, do you obtain the same result regardless of which is performed first? If it matters, which activity should be done first?

a. Drain the used oil from the car's engine.
b. Fill the crankcase with four quarts of new oil.

a. Scrape the dinner dishes.
b. Wash the dinner dishes.

a. Wash your face.
b. Brush your teeth.

About Words

A *commuter* goes back and forth to work. In the morning, she goes to work. At the end of the day, she reverses her steps to return home. The *Commutative Law* tells us when we can reverse the order in which we combine two numbers.

A mathematical law such as $a + b = b + a$ is true for *all* values of a and b. Of course, we cannot check all possible values of a and b to see if they work. However, if we find at least one set of values for the variables that make an equation *un*true, we can be sure the equation is not a law.

Your teacher will give you a copy of Display 2.18 to fill in.

a	b	$a + b$	$b + a$	$a - b$	$b - a$	$a * b$	$b * a$	$a \div b$	$b \div a$
1	2	_____	_____	_____	_____	_____	_____	_____	_____
2	2	_____	_____	_____	_____	_____	_____	_____	_____
3	3	_____	_____	_____	_____	_____	_____	_____	_____
4	3	_____	_____	_____	_____	_____	_____	_____	_____
4	4	_____	_____	_____	_____	_____	_____	_____	_____
1.5	3	_____	_____	_____	_____	_____	_____	_____	_____
6	–3	_____	_____	_____	_____	_____	_____	_____	_____
5	5	_____	_____	_____	_____	_____	_____	_____	_____
5	10	_____	_____	_____	_____	_____	_____	_____	_____

Display 2.18

1. Which parts of the table illustrate the Commutative Law of Addition? Explain.

2. Which parts of the table illustrate the Commutative Law of Multiplication? Explain.

3. For what values of a and b does $a - b = b - a$? Can you find a pattern? Explain.

4. Is there a Commutative Law for subtraction? Why or why not?

5. For what values of a and b does $b \div a = a \div b$? Can you find a pattern? Explain.

6. Is there a Commutative Law for division? Why or why not?

Here are some other laws of arithmetic:

Associative Law of Addition	$a + (b + c) = (a + b) + c$
Associative Law of Multiplication	$a * (b * c) = (a * b) * c$

These equations are abbreviations for statements of the laws. Parentheses are used in the algebraic statement of these laws to tell us what to do first. For example, in

$$19 + (6 + 4) = 19 + 10 = 29$$

we added the 6 and 4 first because 6 + 4 is in parentheses. The Associative Law of Addition tells us that this gives the same result as

$$(19 + 6) + 4 = 25 + 4 = 29$$

where we added the 19 and 6 first. We can write this out in three steps.

$$19 + (6 + 4) = (19 + 6) + 4$$
$$19 + 10 = 25 + 4$$
$$29 = 29$$

a

Give three more examples of the Associative Law of Addition. Show the three steps for each example.

Give three examples of the Associative Law of Multiplication. Show the three steps for each example.

Since $19 + (6 + 4)$ and $(19 + 6) + 4$ represent the same number, we may as well leave out the parentheses. In other situations, parentheses may be necessary.

b

1. Is there an Associative Law for subtraction? Why or why not? How would you interpret

$$4 - 4 - 2$$

2. Is there an Associative Law for division? Why or why not? How would you interpret

$$4 \div 4 \div 2$$

When an expression contains more than one operation symbol, it may be ambiguous. In an earlier section we calculated $T = 0.09 * u + 8.50$. Would this be the same as $T = 8.50 + 0.09 * u$?

We have seen that when we combine more than two numbers, the order in which we carry out the operations may affect the answer. When we write down an expression to be calculated, we can use parentheses to indicate which order we intend. If we do not do that, we may get into trouble.

Ben was absent when his class worked with the electric rate example. He borrowed some notes from a friend and tried to complete the work at home. He does not have a computer or graphing calculator, so he borrowed his mom's four-function calculator to fill in Display 2.10. Here is what Ben turned in.

Usage	Total Bill = 8.50 + 0.09 * Usage
375	$3,221.25
410	$3,521.90
225	$1,932.75
280	$2,405.20
416	$3,573.44
290	$2,491.10
180	$1,546.20
240	$2,061.60

Display 2.19

1. Did Ben use the same formula as the rest of the class?

2. Is Ben's formula correct?

3. Are Ben's numbers correct?

4. Would you want Ben to calculate *your* electric bill?

5. What do you think happened here?

Four-function calculators usually do operations in the order in which you key them in. If you key in

$$4 - 4 - 2$$

it will calculate $(4 - 4) - 2 = -2$ rather than $4 - (4 - 2) = 2$. If you key in

$$2 + 3 * 5$$

it will calculate $(2 + 3) * 5 = 5 * 5 = 25$ rather than
$$2 + (3 * 5) = 2 + 15 = 17$$

a

How could this account for Ben's results? Is there any way Ben could have completed his homework correctly using his mom's calculator?

About Words

The word *convention* comes from Latin words meaning *to come together*. At a political convention, people come together from all over to choose candidates for office. People can come to other kinds of agreement together. They may all agree to drive on the same side of the highway, or to do their mathematical operations in the same order, or anything else they might find convenient. In fact, the word *convenient* comes from the same root as *convention*.

The evaluation of expressions like

$$0.09 * u + 8.50 \text{ and } 8.50 + 0.09 * u$$

are ambiguous. We get different results, depending on which operation we do first. One way to make clear what we mean is to include parentheses to show which operation we want done first. Another way is to adopt some *conventions* about what will be done first. A **convention** is an agreement. In mathematics, we agree to do things the same way so that if different people evaluate the same expression, they will all get the same result.

Mathematicians have adopted some conventions about the order of operations. First, in situations where the order does not matter, we usually do not use parentheses.

Since $a * (b * c) = (a * b) * c$, we usually just write $a * b * c$. Second, if only one operation is involved, we proceed from left to right. We know that $a - (b - c)$ does not equal $(a - b) - c$. The left-to-right convention tells us to interpret $a - b - c$ as $(a - b) - c$. Thus $4 - 4 - 2$ is interpreted as $(4 - 4) - 2 = 0 - 2 = -2$ rather than as $4 - (4 - 2) = 4 - 2 = 2$.

b

Evaluate the following expressions:

$$4 - 2 - 5$$
$$4 + 2 + 5$$
$$4 \div 2 \div 5$$
$$4 * 2 * 5$$

1. Which of the expressions can be evaluated to more than one final value if we follow the laws and conventions of mathematics that we have studied?

2. For each expression with a single correct value, list:

    (a) the correct value

    (b) the law (if any) that leads to that value

    (c) the conventions (if any) that lead to that value.

When we have more than one operation, multiplication and division are done before addition and subtraction. For example:

$$8.50 + 0.09 * 50 \text{ is interpreted as}$$
$$8.50 + (0.09 * 50) = 8.50 + 4.50 = 13$$

Note that this is *not* left-to-right order, which would be

$$(8.50 + 0.09) * 50 = 8.59 * 50 = 429.5$$

Generally, we work from left-to-right doing multiplication and division first, and then return to go from left to right doing addition and subtraction. Most scientific and graphing calculators, and most spreadsheets and programming languages, follow this convention. Be careful when using four-function calculators; many of them evaluate expressions from left to right. If we want to depart from the standard order of operations, we use parentheses to indicate what to do first. When we use parentheses to group items, they are called grouping symbols.

Compute the following. Write down all the intermediate steps. Is the result the same if you ignore the parentheses?

$$(4 + 5) * (7 + 3) - 2$$

$$4 * (5 - 7) * (3 + 2)$$

$$(4 * 5) - 7 + 3 * 2$$

$$2 + 4 * 5 - (7 - 3)$$

$$(4 * 5) - 3 + (2 - 7)$$

At this point we have discussed four laws and several conventions of mathematics.

What are the four laws? Give their full names and state them, using algebraic symbols. How many different operations are involved in each of the four laws if you consider them one law at a time?

The laws we have seen so far deal with a single operation. There is a law of arithmetic that relates two operations. It is the *Distributive Law.*

a

Anindya and Darby had lunch at the Common Man in Ashland, New Hampshire, the other day. Anindya's bill was $5.40, and Darby's bill was $6.20. They agreed to leave a 20% tip. Anindya calculated the total bill and took 20% of that. Darby took 20% of his bill and 20% of Anindya's bill and added them to determine the tip.

1. What tip did Anindya compute?

2. What was Darby's?

3. Did they get the same answer? Will that happen the next time they go to the restaurant?

The **Distributive Law for multiplication over addition** is

$$a * (b + c) = a * b + a * c \qquad (1)$$

That version is relevant to Anindya and Darby because

$$0.20 * (5.40 + 6.20) = 0.20 * 5.40 + 0.20 * 6.20 \qquad (2)$$

b

Answer the following questions about the equations labeled (1) and (2).

1. Match up the calculation in (2) with the Distributive Law in (1). In particular, what are the values of *a*, *b*, and *c*?

2. Which side of the equation (2) corresponds to the calculation Anindya did?

3. Which side of the equation (2) corresponds to the calculation Darby did?

4. What does the Distributive Law (1) say about their results?

5. How many additions did Anindya have to do using her method? How many multiplications?

6. How many additions did Darby have to do using his method? How many multiplications?

7. Do you think one way of doing the calculation is easier than the other? Imagine that they did not bring a calculator to lunch. Why or why not?

This example illustrates one of the main applications of algebra. It enables us to transform one expression into another one that is simpler or more appropriate for some particular purpose.

At the end of the school year, 12 people go to the Common Man for lunch. They are feeling in good spirits and want to leave a 27% tip this time. And not one of them brought a calculator!

1. Which do you think will be the *easier* way to figure the total tip?

   (a) Find the total bill and take 27% of that?

   (b) Find 27% of each bill and total these tips?

   Explain your reasoning.

2. Which would be the *correct* way to figure the total tip? Why?

There are many applications of the Distributive Law. The example with 12 diners used

$$x * (a + b + c + d + e + f + g + h + i + j + k + l) =$$
$$x * a + x * b + x * c + x * d + x * e + x * f + x * g +$$
$$x * h + x * i + x * j + x * k + x * l$$

Show the Distributive Law using different operations.

Here is another. Starting from

$$a * (b + c) = a * b + a * c$$

we can use the Commutative Law of Multiplication three times

$$a * (b + c) = (b + c) * a$$

$$a * b = b * a$$

and

$$a * c = c * a$$

Replacing everything in the original law by an equal quantity, we get

$$(b + c) * a = b * a + c * a$$

There are applications of the Distributive Law involving subtraction, such as

$$a * (b - c) = a * b - a * c$$

and

$$(b - c) * a = b * a - c * a$$

You need to be able to recognize and use all these different forms.

Problem Set: 2.3

1. Determine whether or not the two activities are commutative, that is, do you obtain the same result regardless of which is performed first? If it matters, which activity should be done first?

   (a) Set the table.
   Serve the meal.

   (b) Fill the tub with water.
   Take a bath.

   (c) Clean the house.
   Rake the leaves.

   (d) Put on your shoes.
   Put on your socks.

   (e) Strap on your parachute.
   Jump out of the airplane.

   (f) Do your history homework.
   Do your math homework.

   (g) Do all the multiplications and divisions in an expression.
   Do all the additions and subtractions in an expression.

2. Here are some examples of the laws we have studied in this section. For each, give the law (or laws) it is an example of. Then do the calculation, writing down intermediate steps to show that each equation is true.

   (a) $(3 + 7) + 9 = 3 + (7 + 9)$
   (b) $(3 + 7) + 9 = (7 + 3) + 9$
   (c) $(3 + 7) + 9 = 9 + (3 + 7)$
   (d) $(3 + 7) + 9 = 9 + (7 + 3)$
   (e) $(3 + 7) + 9 = 3 + (9 + 7)$
   (f) $(3 + 7) + 9 = (9 + 7) + 3$

   (g) What did you notice about adding three numbers in (a) through (f)? Will this work for adding more than three numbers? Write a paragraph discussing how the Commutative and Associative Laws of Addition support your conclusion.

3. Here are some examples of the laws we have studied in this section. For each, give the law (or laws) it is an example of. Then do the calculation, writing down intermediate steps to show that each equation is true.

   (a) $(3 * 7) * 9 = 3 * (7 * 9)$

   (b) $(3 * 7) * 9 = (7 * 3) * 9$

   (c) $(3 * 7) * 9 = 9 * (3 * 7)$

   (d) $(3 * 7) * 9 = 9 * (7 * 3)$

   (e) $(3 * 7) * 9 = 3 * (9 * 7)$

   (f) $(3 * 7) * 9 = (9 * 7) * 3$

   (g) What did you notice about multiplying three numbers in (a) through (f)? Will this work for multiplying more than three numbers? Write a paragraph discussing how the Commutative and Associative Laws of Multiplication support your conclusion.

4. Compute the following:

   (a) $5 * 4 - 2 * 5 + 6 \div 3$

   (b) $5 + 6 * 3 - 7$

   (c) $15 - 9 * 2 \div 3$

   (d) $3 * (14 + 2)$

   (e) $18 + 6 * (1\frac{1}{2} + \frac{1}{2})$

   (f) $2 * (8 - 5) + 32 \div 4$

   (g) $6 * 7 - 5 + 4 * 3$

   (h) $13 - 6 * 2$

   (i) $18 - 4 \div 2 * 5$

   (j) $5 * (3 - 2) + 6 \div 3$

   (k) $7 \div (5 - 3) + 2$

   (l) $5 - 3 * 8 \div 6$

5. Tauheedah was referred to as "The Human Calculator" by her classmates because she could very easily add and multiply numbers in her head. When asked what her secret was, she said, "It's easy, just use the Commutative and Associative Laws of Addition and Multiplication."

Here's an example of her technique:

$$15 + 7 + 5 + 3 = 15 + 5 + 7 + 3 = (15 + 5) + (7 + 3) = 20 + 10 = 30$$

Use the Commutative and Associative Laws to do the following arithmetic problems mentally. Write down how you grouped the numbers and what laws you used.

(a) $24 + 28 + 12 + 46$

(b) $35 + 29 + 34 + 71$

(c) $7.4 + 89 + 13.6$

(d) $(23.4 + 8.7) + 5.6$

(e) $88 * 2 * 5$

(f) $25 * 19 * 4$

(g) $0.25 * 0.9 * 4$

(h) $5\% * 0.7 * 0.8 * 20$

6. Compute the following. Write down all intermediate steps.

(a) $(4 + 5) + (7 + 3) + 2$

(b) $4 + (5 + 7) + (3 + 2)$

(c) $(4 + 5) + 7 + (3 + 2)$

(d) $2 + (4 + 5) + (7 + 3)$

(e) $(4 + 5) + 3 + (2 + 7)$

(f) What did you notice about adding five numbers in (a) through (e)? Will this work for adding more than five numbers? Write a paragraph discussing how the Commutative and Associative Laws of Addition support your conclusion.

7. Compute the following. Write down all intermediate steps.

   (a) $(4 * 5) * (7 * 3) * 2$
   (b) $4 * (5 * 7) * (3 * 2)$
   (c) $(4 * 5) * 7 * (3 * 2)$
   (d) $2 * (4 * 5) * (7 * 3)$
   (e) $(4 * 5) * 3 * (2 * 7)$

   (f) What did you notice about multiplying five numbers in (a) through (e)? Will this work for multiplying more than five numbers? Write a paragraph discussing how the Commutative and Associative Laws of Multiplication support your conclusion.

8. Tauheedah had another method for multiplying numbers in her head that used the Distributive Law. Here are examples of her method:

   $$99 * 58 = (100 - 1) * 58 = 100 * 58 - 1 * 58 = 5800 - 58 = 5742$$

   $$12 * 8.05 = 12 * (8 + 0.05) = 12 * 8 + 12 * 0.05 = 96 + 0.60 = 96.60$$

   Do the following arithmetic mentally by using the Distributive Law. In each case, give an algebraic statement of the pattern you are using.

   (a) $67 * 99$
   (b) $98 * 38$
   (c) $103 * 30$
   (d) $49 * 25$
   (e) $836 * 45 + 164 * 45$
   (f) $4 * 12.25$
   (g) $8 * 25.25$
   (h) $12 * 5.95$
   (i) $15 * 3.98$
   (j) $59 * 2.75 - 19 * 2.75$
   (k) $15 * 3\frac{1}{3}$
   (l) $20 * 5\frac{9}{10}$
   (m) $\frac{1}{5} * 25\frac{1}{3}$
   (n) $36 * 3\frac{11}{12}$

9. Henry bought three lunch orders, one for himself and two for friends. Each order consisted of a cheeseburger costing $1.69, a large fries costing $1.29, and a medium cola costing $1.09. How can Henry find the total cost with only one multiplication? What was the total cost? Which of the laws you studied in this section are involved here?

10. Lisa was in charge of setting up the chairs for her Girl Scout Troop's play presentation. She set up three sections of chairs. Two of the sections had six rows with eleven chairs in each row, and one section had five rows with twelve chairs in each row. Draw a diagram showing how Lisa might have set up the chairs. Write a numerical expression underneath the diagram showing how many chairs are in each section. How many chairs did Lisa set up in all?

11. The Whatta Gas station sells three grades of fuel: regular, super, and premium. The price per gallon is $1.939 for regular, $2.039 for super, and $2.139 for premium. Select variables to represent the amounts of the three kinds of gasoline sold. Write an expression for the total amount of money received for all three kinds of gasoline in terms of the number of gallons of each one sold. Then use the expression to find out how much money the station takes in on a day when it sells 12,000 gallons of regular, 8000 gallons of super, and 5000 gallons of premium. Write a paragraph about what this has to do with "order of operations."

2.4 Solving Equations

Alicia's family lives in Houston, Texas. Because of the warm weather in that part of the country, her family installed a window air conditioner on July 1. Alicia is curious to know how much electricity the air conditioner uses. She decided to compare the number of kiloWatt-hours of electricity they used this July (with the air conditioner) with the number of kiloWatt-hours used last July (without the air conditioner).

> **Why do you think that Alicia decided to compare the amount of electricity used this July with the amount used last July? Do you think that it would make more sense to compare the amount of electricity used this July with the amount used this June? Why or why not?**
>
> **Will the comparison of this July with last July accurately answer Alicia's question? Why or why not?**

a

Learning Outcomes

After studying this section, you will be able to:

Determine whether a particular number is a solution for an equation

Identify some equations that have no solutions and some equations that have more than one solution

Explain the meaning of algorithm and give examples of algorithms

Describe and use an algorithm for solving an equation of the form $y = ax + b$, for x, given numerical values for a, b, and y.

Alicia knows that the easiest way to compare one month with another is to read the number of kiloWatt-hours used directly from the electric bill. She has this month's bill, which says that they used 450 kWh. Unfortunately, she does not have a copy of last July's bill. She has asked her father to look back in his checkbook and find the amount they paid for electricity last July. She finds that they paid $46.75 for electricity last July. Alicia begins with this amount and retraces the electric company's steps in calculating the bill.

> **What *are* the steps in calculating an electric bill? How would you calculate the bill for *this* July, when 450 kWh were used?**

b

To undo the step of adding the $8.50 customer service charge, Alicia subtracts $8.50. This gives her

$$\$46.75 - \$8.50 = \$38.25$$

How many kiloWatt-hours of electricity did they use last July? To find this out, Alicia retraces the electric company's steps again. She knows that $38.25 is $0.09 times the number of kiloWatt-hours used. To undo multiplying the number of kiloWatt-hours used by $0.09, Alicia divides $38.25 by $0.09. This gives her the number of kiloWatt-hours used last July.

$$38.25 \text{ dollars} \div 0.09 \, \frac{\text{dollars}}{\text{kWh}} = 425 \text{ kWh}$$

Alicia's reasoning could be applied to *any* past bill.

How much did the energy use increase from last July to this July? Do you think that this increase is completely due to the air conditioner, or could there be other reasons? Explain your answers.

Find the number of kiloWatt-hours of electricity used in the months that have these bills,

$35.50, $19.93, $82.39

The electric company billing policy can be described by the equation

$$T = 0.09 * u + 8.50$$

For Alicia's electric bill, we knew $T = \$46.75$, so we replaced T with 46.75. We reasoned that $u = 425$ kWh in that case. We call this process *solving the equation*.

$$46.75 = 0.09 * u + 8.50$$

About Words

An Arabic book about solving equations, *Hisab al-jabr w'al muqabalah*, was translated into Latin and became very well known in 13th century Europe. The Latin form of *al-jabr* became *algebra*, and gradually this became a convenient way to refer to the study of equations and related ideas.

A Word to Know: A solution to an equation in one variable is a value for the variable that makes the equation true.

If we try replacing u with 425 in $0.09 * u + 8.50$, we get $0.09 * 425 + 8.50$.

Since $0.09 * 425 + 8.50 = 38.25 + 8.50 = 46.75$, we know that $46.75 = 0.09 * u + 8.50$ is a true equation when $u = 425$. This calculation checks our reasoning process and shows that 425 really is a solution to the equation.

If we had made a mistake and concluded that $u = 400$, replacing u by 400 in $0.09 * u + 8.50$ would give

$$0.09 * 400 + 8.50 = 36.00 + 8.50 = 44.50$$

Since this is *not* equal to 46.75, we know that $u = 400$ is not a solution to the equation $46.75 = 0.09 * u + 8.50$. Any time we think we have a number that is a solution to an equation, we have to try the number in the equation to see if it really works as a solution.

Alicia found the electrical usage for three more months. Check her work in Display 2.20.

a

Bill, $	Usage, kWh
66.28	640
44.50	400
60.07	573

Display 2.20

Each equation we have solved has had only one solution. In other situations we may find several solutions to an equation, or none at all. To deal with this, we introduce the idea of a *solution set* for an equation.

A Phrase to Know: A solution set is the set of *all* the numbers that are solutions to the equation. The equation

$$46.75 = 0.09 * u + 8.50$$

has just one solution, $u = 425$, so we would write the solution set as {425}.

Here is another example of an equation:

$$0 * k = 37$$

What is your favorite number? Replace *k* in our equation with your favorite number, and see if you get a true statement. Try your second favorite number, too.

b

No matter what number we replace k with, we get $0 * k = 0$, not 37, so there is no solution to this equation. We write the solution set as { } to represent the **empty set**, the set with no elements.

What are the solutions of $x = x + 1$?

Here is an example of an equation with more than one solution:

$$(a - 1) * (a - 3) = 0$$

If we replace a by 1, we get

$$(1 - 1) * (1 - 3) = 0 * (1 - 3) = 0$$

so 1 is a solution. If we replace a with 3, we get

$$(3 - 1) * (3 - 3) = (3 - 1) * 0 = 0$$

so 3 is a solution, as well. If these are the only solutions (and they are), we write the solution set as {1, 3}

 All solutions to

$$(x - 2) * (x - 4) = 0$$

are in the list 1, 2, 3, 4, 5. Check to see which of 1, 2, 3, 4, and 5 actually are solutions to the equation.

Write the solution set.

To solve equations more quickly, we need a step-by-step procedure to follow. Such a procedure is called an **algorithm.** A recipe for making brownies or instructions for rebuilding a carburetor are good examples of algorithms. Once we have an algorithm, we can just follow the steps. An advantage is that this lets us work much faster. A disadvantage is that we may not think about what we are doing.

The equation we have been working with is

$$T = 0.09 * u + 8.50$$

In the most recent examples, we have had a value for T and wanted to know u. We will now study an algorithm for finding u in this equation. The algorithm is based on taking the equation and calculating with it. The steps are based on *laws governing equality.* Here are some of those laws:

1. If two quantities are equal and we add the same number to both, then the results will be equal.

   For example, suppose Ted and Fred were the same height at the end of the last school year. If each boy grew two inches over the summer, then Fred would still be the same height as Ted. We can abbreviate the general principle with algebra.

   $$\text{If } a = b, \text{ then } a + c = b + c$$

2. If two quantities are equal and we subtract the same number from both, then the results will be equal.

   For example, if Sarah and Saul had the same "best ever" golf score at the end of the last school year, and each lowered their best score by three shots over the summer, they would still have the same "best ever" golf score. We can abbreviate the general principle with algebra.

   $$\text{If } a = b, \text{ then } a - c = b - c$$

3. If two quantities are equal and we multiply both by the same number, then the results will be equal.

   For example, if Juanita and Soo are the same height in inches, and we convert each of their heights to centimeters by multiplying by 2.54, then they would have the same height in centimeters. We can abbreviate the general principle with algebra.

   $$\text{If } a = b, \text{ then } a * c = b * c$$

4. If two quantities are equal and we divide both by the same number, then the results will be equal, *provided we do not divide by zero*. Division by zero is not defined.

   For example, if four friends split a dinner check of $24 equally on Monday night by dividing the dinner check by four, and they get a check for $24 for Friday's dinner, the shares will be the same as on Monday. We can abbreviate the general pattern with algebra.

   $$\text{If } a = b \text{ and } c \text{ is not } 0, \text{ then } a \div c = b \div c$$

   Why must c not be 0?

   We can apply these principles to our equation

   $$T = 0.09 * u + 8.50$$

   Since T and $0.09 * u + 8.50$ are equal, we can use any of the four principles. In one example we did earlier, T was $42.25, so the equation we wanted to solve was

   $$42.25 = 0.09 * u + 8.50$$

   Our algorithm says to start by subtracting 8.50 from each of the two equal quantities, T and $0.09 * u + 8.50$, in order to undo the addition. If we subtract 8.50 from 42.25, it gives us

   $$42.25 - 8.50 = 33.75$$

If we subtract 8.50 from $0.09 * u + 8.50$, the result is

$$0.09 * u$$

Subtracting 8.50 undoes adding 8.50. For this reason, addition and subtraction are called *inverse operations*. After we subtract the 8.50, we have the new equation

$$33.75 = 0.09 * u$$

Now we divide the equal quantities 33.75 and $0.09 * u$ by 0.09. If we divide $0.09 * u$ by 0.09 we get u. Dividing by 0.09 undoes the effect of multiplying by 0.09. Multiplication and division are *inverse operations*.

When we divide 33.75 by 0.09, we get 375. Now we have the new equality

$$375 = u$$

which is the value of u when $T = 42.25$. We say that the original equation has been **solved for** u. The value that we found for u should be checked in the original equation

$$0.09 * (375) + 8.50 = 33.75 + 8.50 = 42.25$$

Notice that we could write $375 = u$ as $u = 375$. Some people prefer to write $u = 375$ because this is a statement about the value of u, and the subject of a sentence usually comes first in English.

Display 2.21 lists the steps for the algorithm for solving the equation: $42.25 = 0.09 * u + 8.50$. Copy the table, and write a comment about why each step is done and the mathematical principle used.

Steps in Algorithm	Comment and Mathematical Principle
$42.25 = 0.09 * u + 8.50$	
$42.25 - 8.50 = 0.09 * u + 8.50 - 8.50$	
$33.75 = 0.09 * u$	
$33.75 \div 0.09 = 0.09 * u \div 0.09$	
$375 = u$	

Display 2.21

Now it's time for you to solve some equations. Be sure to check your answers in the original equation.

a

1. Solve $37.75 = 0.09 * u + 8.50$ for *u*.

2. Solve $26.50 = 0.09 * u + 8.50$ for *u*.

3. Solve $42.10 = 0.09 * u + 8.50$ for *u*.

4. Solve $31.40 = 0.09 * u + 8.50$ for *u*.

Suppose you encounter an equation like

$$18 = 3 * (x + 2)$$

You can use the Distributive Law to replace $3 * (x + 2)$ with $3x + 6$. The new equation

$$18 = 3x + 6$$

is one you know how to solve.

Use this method to solve the following equations for *x*:

b

1. $27 = 3 * (x + 2)$

2. $19 = 4 * (x + 7)$

You can see that the process of solving an equation requires us to write *u* over and over again. For pencil and paper calculations, one letter abbreviations save a lot of writing. In computer programming, where the machine does all the calculating, programmers often use longer abbreviations, such as "kWh" or "Total". This helps them to remember what the original variables represent. It is traditional in algebra to omit a multiplication sign between a number and a letter. For example, $0.09 * u$ would probably be written $0.09u$. In entering this into a computer, you may need to type the $*$.

Why can't we omit the multiplication sign between two numbers?

Problem Set: 2.4

1. Suppose $26.25 = 2.7t + 6$.

    (a) Is 2.7 a solution to this equation? How do you know?

    (b) Is 0 a solution to this equation? How do you know?

    (c) Is 6 a solution to this equation? How do you know?

    (d) What is meant by the solution to this equation?

    (e) Find a solution for this equation.

    (f) Explain how you arrived at the solution.

2. Suppose $5 = 0 * x - 3$.

    (a) Is 0 a solution to this equation? How do you know?

    (b) What is the solution set for this equation? Explain how you arrived at your answer.

3. Suppose $5 * x = 5 * x$.

    (a) Is 0 a solution to this equation? How do you know?

    (b) Is −8 a solution to this equation?

    (c) Is 2.3 a solution to this equation?

    (d) What can you say about the solution set to this equation?

4. Recall that x^2 is an abbreviation for $x * x$.

    (a) The equation $x^2 = 16$ has two solutions. Find them and write the solution set.

    (b) How many solutions does $x^2 = 49$ have? Write the solution set.

    (c) How many solutions does $x^2 = -4$ have? Write the solution set.

    (d) How many solutions does $x^2 = 0$ have? Write the solution set.

5. Solve $45 = 2 * u + 15$ for u. Write an explanation for your steps.

6. The Cheapo Rental Agency charges $90 a week and 35 cents a mile to rent a Taurus.

   (a) Write an equation giving the total cost for a week. Be sure to write down what your variables represent.

   (b) Use your equation to answer these questions.

      (i) You drive 250 miles during the week. What is your total cost?

      (ii) Cheapo sends you a bill for $212.50. How many miles do they claim you drove?

   (c) The Bargain Rental Agency charges $142 a week and 10 cents a mile to rent a Taurus.

      (i) Write an equation giving the total cost for a week. Be sure to write down what your variables represent.

      (ii) If you drive 250 miles during the week, what is your total cost for the Taurus?

      (iii) How many miles would you have to drive for the cost to be $212.50?

   (d) Which agency has the better deal? Explain your answer.

7. Solve $35.2 = 0.5u + 6.7$ for u. Write an explanation for your steps.

8. Solve $582 = 7x + 15$ for x. Write an explanation for your steps.

9. Solve $14 = 10y - 8$ for y. Write an explanation for your steps.

10. Questions 6–9 above were very similar to the electric company example. The equations below are slight variations, but you can still solve for x. In each case, explain what made the problem a little different.

    (a) $18 = 10 + 2x$

    (b) $14 = 10x - 8$

    (c) $3x + 10 = 2$

    (d) $3x - 10 = 2$

11. Use your calculator to solve for x.

(a) $21.35 = 5.77 + 3.12x$

(b) $14{,}583 = 5.77x - 8.4$

(c) $4.09x + 5.77 = 3.12$

(d) $4.09x - 5.77 = 3.12$

12. Solve for x.

(a) $x^2 = 144$

(b) $(x - 5) * (x + 4) = 0$

(c) $(4 + x) * (x + 3) = 0$

(d) $(x + 7)^2 = 0$

(e) $(x - 4.87) * (3.483 + x) = 0$

(f) $x^2 = 2$

13. Solve $u + a = a$ for u.

14. Solve for y.

(a) $5y = 20$

(b) $-3y = 87$

(c) $14 = y + 9$

(d) $y - 3 = 57$

(e) $14 = 2y$

Solve problems 15–16.

- If you use algebra to solve the problem, identify any variables.

- If you use a combination of reasoning and arithmetic, explain the process you used.

15. Nathan bought a 120 square-foot rug on sale for $3.85 a square foot. If he only used 104 square feet, what area of the rug was left over? How much had he paid for the leftover part?

16. Raquel has $294 to buy a Rockhopper bicycle. If the bicycle costs $440, how much more money does she need to buy the bicycle?

17. Stop and Shop sells lean ground beef that has approximately 10% fat and regular ground beef that has approximately 17% fat.

  (a) If you make hamburgers that are a half pound each, approximately how much fat is in one burger made with lean ground beef?

  (b) Using L to represent the number of pounds of lean ground beef and F to represent the number of pounds of fat in it, write an equation for the amount of fat in L pounds of lean ground beef. Use your equation to determine how much fat would be in 2.5 pounds of lean ground beef.

  (c) Let G represent the number of pounds of regular ground beef. Write an equation for the amount of fat in G pounds of regular ground beef. Use your equation to calculate how much fat is in 1.5 pounds of regular ground beef.

  (d) Suppose you want to mix the two kinds of ground beef. Write an equation for the total fat in a mixture of L pounds of lean ground beef and G pounds of regular ground beef. What is the total fat in a mixture of 3 pounds of lean ground beef and 2 pounds of regular ground beef?

18. The Drama Club is doing a production of *Grease* this summer. Rental of the costumes and the auditorium will cost approximately $1,200. Adult tickets will sell for $5, and student/senior citizen tickets will sell for $3.50.

  (a) Choose appropriate letters for the variables in this situation.

  (b) Write down what each variable represents.

  (c) Write an equation that could be used to determine the amount of money the Drama Club collects from those who attend.

  (d) If 90 students, 10 senior citizens, and 50 adults attend the play, how much money will the Drama Club collect?

  (e) Use your equation to find the money collected if 208 adults and 156 students attend the production.

(f) If the rental costs are their only costs, write an equation that could be used to determine the profit the Drama Club makes on their production. (*Profit* is the money collected, minus the money paid out in costs; that is, it's the amount of money you come out ahead.)

(g) Use your equation to find the profit if 218 adults and 256 students attend the production.

(h) The play ran Friday and Saturday nights. Use your equations to find the money collected and the profit if 117 students and 104 adults attended Friday night, and 160 students, 16 senior citizens, and 67 adults attended Saturday night.

19. Solve for x.

(a) $3 * (7 + x) = 21$

(b) $3 * (7 - x) = 21$

(c) $27 = 3 * (x - 2)$

20. Earlier you used the Distributive Law to solve

$$18 = 3 * (x + 2)$$

You could also divide both sides of this equation by 3 to get

$$6 = x + 2$$

also an equation you know how to solve. Use this algorithm to solve the same equations you solved in problem 19. Then write a paragraph evaluating the pros and cons of the two algorithms.

2.5 A New Pattern

Alfredo had a birthday last June. His Aunt Mercedes gave him $1,000. He was going to spend it on video cartridges, but his aunt offered him a deal. She will hold the money for him and send him a check for 7% interest on the $1,000 for each year she keeps it. Or he can collect the $1,000 and the interest earned whenever he wants. Once he collects it, though, he will not earn any more interest.

What is 7% of $1,000? How do you compute that? What is 7% of $1,295?

Learning Outcomes

After studying this section, you will be able to:

Explain how whole number exponents work

Evaluate expressions that contain whole number exponents

Use exponentiation with other operations of arithmetic.

Alfredo is just starting school. He has dreamed of visiting his cousin Alphonse in Central America after graduation. He makes a table showing what happens if he leaves the money with Aunt Mercedes for four years.

Copy Display 2.22 showing Alfredo's finances. Fill in the missing values.

After this many years	Amount left with Aunt	Interest Earned	
		This year	To date
0	$1,000	$0	$0
1	$1,000	$70	$70
2	$1,000	$70	$140
3			
4			

Display 2.22

If Alfredo collects the gift from his aunt after four years, what is the total amount of money he will have received from her?

Alfredo shows his calculations to his aunt. She tells him that what he calculated is called *simple interest*. **Simple interest** is based on the original amount, so the interest is the same every year. Aunt Mercedes has many investments that pay simple interest. Every year she receives an interest check from each investment. She uses the money to do things such as buying presents for her favorite nephew.

Alfredo decides that he would rather save his interest for his trip. Aunt Mercedes says that if he does that, she will pay him interest on the interest! This is called **compound interest**. It means that after a year Alfredo has $1,070 and the second year interest will be 7% of $1,070 instead of 7% of $1,000.

What is 7% of $1,070? How much will Alfredo have after two years at a compound interest rate? Copy and complete the table in Display 2.23.

	Simple Interest			Compound Interest	
After this Many Years	Interest this Year	Total from Aunt		Interest this Year	Total from Aunt
0	$0	$1,000		$0	$1,000
1	$70	$1,070		$70	$1,070
2	$70	$1,140		$74.90	$1,144.90
3					
4					

Display 2.23

You can see that Alfredo makes more with compound interest. One way to compare simple interest with compound interest is to compute the average amount of interest for each over the time period. To do this for the four years above

1. Subtract the initial amount of $1,000 from how much money Alfredo has at the end of four years. This gives the total interest earned. For simple interest, this is $280.

2. Divide this amount by 4. This gives the average amount of interest per year.

1. Find the average interest on Alfredo's savings for the four year period described, at simple interest. Do you get the same result as if you had averaged the interest for years 1 through 4?

2. Do the same for compound interest.

3. You know how much Alfredo had after four years at 7% compound interest. What simple interest would he have to earn to make the same amount after four years?

Alfredo estimates the cost of visiting his cousin to be $2,400. Even at compound interest, he will not have enough money after four years. Perhaps he should wait longer. He could make his trip while he is in college, or after he graduates from college.

Continue Alfredo's financial analysis through his college years.

1. Figure out how much money Alfredo will have after 5, 6, 7, and 8 years at compound interest. Add these results to the table you made earlier (Display 2.23).

2. Figure out how much money Alfredo will have after 5, 6, 7, and 8 years at simple interest. Add these results to your table.

3. What do you think the average interest on Alfredo's savings for years 5–8 will be at simple interest? What about between years 0 and 8? Check both your estimates with your calculator.

4. What do you think the average interest on Alfredo's savings for years 5–8 will be at compound interest? What about between years 0 and 8? Check both your estimates with your calculator.

It looks like eight years is still not enough for Alfredo to have $2,400. He is curious how long that will take.

Do you think Alfredo will *ever* receive $2,400 from his aunt's gift?

So far we have been working out a table for Alfredo's finances one year at a time. We calculated the amount for each year based on the amount for the previous year. A method that does this is called a **recursive algorithm.** This is a good method if we want an entire table. It is not a very good method if we want to find out how long it will take until Alfredo has $2,400.

Lorna says that instead of adding 0.07 of last year's amount to this year's amount, we could just multiply last year's amount by 1.07. That is, she wants to calculate the amount Alfredo has after two years as

$$1.07 * 1070$$

rather than

$$1070 + 0.07 * 1070$$

Does this give the correct answer?
Does $a + 0.07a = 1.07 * a$ for any value of a? Try different values of a and discuss your answers.
Does it matter which method we use?
Carefully count how many buttons you have to press to calculate

$$1070 + 0.07 * 1070$$

and record the result. Then count how many buttons you have to press to calculate

$$1.07 * 1070$$

Which procedure is faster? Does it save much time if you want to calculate just one additional year? What if you want to make a table covering 25 years?

Copy and complete Display 2.24 based on 7% compound interest. Use Lorna's shortcut.

After this Many Years	Amount of Money from Aunt
0	$1,000
1	
2	
3	
4	
5	
6	
7	
8	
9	
10	

Display 2.24

We can use a pattern in our table to do this even more quickly. After two years, Alfredo has

$$\$1,070 * 1.07 = \$1,144.90$$

Where did we get that $1,070 from? It was just $1,000 * 1.07. We can replace $1,070 in the previous equation with the equal quantity $1,000 * 1.07 to get

$$(\$1,000 * 1.07) * 1.07 = \$1,144.90$$

the amount after two years.

What law of arithmetic says ($1,000 * 1.07) * 1.07 = $1,000 * (1.07 * 1.07)

a

After three years, he has

$$\$1,144.90 * 1.07$$

and that is the same as

$$\$1,000 * (1.07 * 1.07 * 1.07)$$

What is the pattern here? Can you express how much money Alfredo has saved after six years in terms of just the starting amount, $1,000, and the multiplier 1.07?

b

There is an abbreviation for expressions like

$$1.07 * 1.07 * 1.07 * 1.07 * 1.07 * 1.07$$

and

$$1.07 * 1.07 * 1.07 * 1.07$$

Actually, there are at least two abbreviations. We can write

$$1.07 * 1.07 * 1.07 * 1.07 * 1.07 * 1.07$$

as 1.07^6

or as $1.07\text{^}6$

We can write

$$1.07 * 1.07 * 1.07 * 1.07$$

as 1.07^4

or as $1.07\text{^}4$

Use the expressions above and the multiplication key on your calculator to find 1.07^4 and 1.07^6.

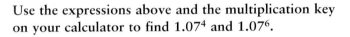

In the expression 1.07^4, the small raised 4 is called an **exponent** and the 1.07 is called the **base**. In this case, the 4 tells us how many 1.07's are to be multiplied together and is an abbreviation for repeated multiplication. The operation of doing this is called **exponentiation.** We would also say we

About Symbols

The caret "^" is sometimes used to represent repeated multiplication, especially on computers and calculators. Early computers had input terminals like a typewriter. They could not type exponents like the 4 in 1.07^4, so they had to find another way to represent repeated multiplication.

raised 1.07 to the fourth **power.** Doing this is an operation on numbers, just as addition, subtraction, multiplication, and division are operations on numbers.

How is multiplication repeated addition?

Here are some examples.

$$4^3 = 4 * 4 * 4 = 64 \qquad 2^5 = 2 * 2 * 2 * 2 * 2 = 32$$

$$5^2 = 5 * 5 = 25$$

Using the other notation, we have

$$4 \wedge 3 = 4 * 4 * 4 = 64 \qquad 2 \wedge 5 = 2 * 2 * 2 * 2 * 2 = 32$$

$$5 \wedge 2 = 5 * 5 = 25$$

The "$\wedge$" notation is more recent. It is most frequently used with calculators and computers. Keyboards often do not provide a way to make numerals smaller or display them above numerals and variables. Your calculator may have a key marked y^x or $\wedge$ for exponentiation.

Use the exponentiation key on your calculator to find 1.07^4 and 1.07^6. Do you get the same result as you did using the multiplication key? Which method would you prefer if you wanted to find 1.07^20? Why?

We can use algebra to abbreviate the compound interest pattern in Alfredo's savings. After n years have gone by, he has $1,000 * 1.07^n$ dollars in the bank.

We can use this expression and a calculator to find out how much money Alfredo has after 20 years (*without having to figure out how much money he has after 19 or 18 or 17 years*).

1. Use this expression to calculate how much Alfredo has after 2, 5, and 10 years.

2. How would you interpret 1.07^1 if the expression is correct for $n = 1$?

3. How would you interpret 1.07^0 if the expression is correct for $n = 0$?

4. How much money does Alfredo have after 20 years? Do you think he will ever have enough money to visit his cousin? Explain.

Compound interest is an example of exponential growth. Something grows exponentially if it increases by a fixed multiple for each time period. For Alfredo, the amount of money he had grew by a multiple of 1.07 each year. After n time periods, it would have grown by a factor of 1.07^n. Hence the name *exponential growth*.

Population often grows exponentially. In Display 2.25 are world population figures from 1950 to 2000.

Year	Population
1950	2,500,000,000
1960	3,050,000,000
1970	3,700,000,000
1980	4,450,000,000
1990	5,250,000,000
2000	6,050,000,000

Display 2.25

Starting from 1950, the population grew by about 20% per decade.

1. Calculate what the population would have been in 1960, 1970, 1980, 1990 and 2000 if the population had grown at exactly 20% per decade.

a

2. How do your calculated figures compare with the true values in Display 2.25?

3. Estimate world population in the years 2010 and 2020.

If the multiple is less than 1, we have **exponential decay**. For example, suppose 1,000,000 people knew how to use a slide rule in 1970, about the time inexpensive calculators first became widely available. If the number of people who knew how to use a slide rule then declined by 5% per year, then in 1971 we would have $100\% - 5\% = 95\%$ as many, or 950,000. In 1972, we would have $1,000,000 * 0.95^2 = 902,500$.

According to this scenario, how many people would know how to use a slide rule in the year 2010? In 2020?

b

Problem Set: 2.5

1. Recall that the area A of a square is expressed as s^2, where s is the length of a side. Find the area when a side is 8 feet long. What are the units of your answer?

2. Recall that the volume V of a cube is expressed as s^3, where s is the length of a side. Find the volume when a side is 8 feet long. What are the units of your answer?

3. Calculate the following:

    (a) 3^4   (b) 4^3   (c) 8^1

    (d) 1^8   (e) 10^4   (f) 10^100

    (g) 1.4^3   (h) 13^0   (i) 2^10

4. Calculate the following:

    (a) 3^2 (b) 4^1 (c) 8^5

    (d) 1^5 (e) 10^{14} (f) 10^{1000}

    (g) 1.4^2 (h) 0.1^5 (i) 2^0

5. Recall that the area A of a circle is expressed as πr^2, where r is the radius. Find the area when the radius is 8 feet.

6. Claire has just started to work for the Chesapeake Company for $18,000 per year. The company has a policy of increasing salaries 5% each year. What will Claire earn next year? The year after? In how many years will she be earning $30,000?

7. What is the initial amount Aunt Mercedes would have to give Alfredo today at 7% compound interest in order for him to have $2,400 for his trip four years from now, when he graduates?

8. Suppose the owner of the Cleveland Indians baseball team offered Juan Gonzalez a choice of two ways to earn bonus money at the beginning of the year. He could receive Offer 1: $5,000 for each home run or Offer 2: 1¢ if he hits one home run, 2¢ if he hits two home runs, 4¢ if he hits three home runs, 8¢ if he hits four home runs, etc. Copy and complete the table in Display 2.26.

    Do you think Offer 1 would ever be better than Offer 2? For how many home runs? As his agent, which offer would you have him accept? Why?

Number of Home Runs	INCOME	
	Offer 1	Offer 2
1	$5,000	$0.01
2	$10,000	$0.02
3	$15,000	$0.04
4	____	____
5	$25,000	____
10	$50,000	____
15	____	$163.84
20	____	____

Display 2.26

9. A photocopier enlarges a picture by 25% so that an 8×10 figure is enlarged to 10×12.5. If this is done to the same picture three times, how large is the final one? How many times as large as the original picture will the final one be? What if we enlarge the picture 10 times?

10. The same photocopier reduces a picture by 25% so that an 8×10 picture is reduced to 6×7.5. If this is done to the same picture four times, how small is the final one? How many times as small as the original picture would the final one be? What if we reduce the picture 10 times?

11. A super ball bounces to 0.9 of the height it started from. How high would it be after its fifth bounce if you dropped it from a height of six feet?

12. At electricity generating plants burning coal or oil, an electrostatic precipitator is used to remove smoke particles from the waste gases. Each stage of the precipitator removes 85% of the smoke particles, leaving behind the other 15%. Two stages leave behind 15% of 15%, or $0.15 * 0.15 = .0225$, or two and one-quarter percent. This means $100\% - 2.25\% = 97.75\%$ of the particles have been removed.

(a) What percentage is left behind by three stages? What percentage is *removed* by three stages?

(b) Use exponents to write an equation for how much is left behind by s stages.

(c) Suppose the EPA (Environmental Protection Agency) says we need to remove 99.44% of the smoke particles. How many stages do we need to do this?

(d) Suppose the EPA says we need to remove 99.99% of the smoke particles. How many stages do we need to do this?

2.6 Growth and Scientific Notation

Learning Outcomes

After studying this section, you will be able to:

Interpret scientific notation in science or on a calculator screen

Use scientific notation to abbreviate very large and very small numbers

Explain how integer exponents are used in real life situations

Evaluate expressions that contain integer exponents.

Computers do a lot of computing, but their main use today is storing and using data. For example, records of your family's electrical usage are probably maintained by the electric company on a computer. Each month, the computer prints out your electric bill. The electric company will get a lower rate on the postage to mail the bills if they sort them according to ZIP code. Computer scientists have studied many different algorithms for sorting data. If you only have a few items to sort, any method will do. If you need to sort a lot of data, then the algorithm you use may make a big difference.

Display 2.27 shows three algebraic expressions and their values, when $x = 0$, 1, and 2. These might represent the time to sort x records.

x	$2 * x$	$2^\wedge x = 2^x$	$x^\wedge 2 = x^2$
0	0	1	0
1	2	2	1
2	4	4	4

Display 2.27

The values of these three expressions are similar when x is a small value.

Copy Display 2.27 onto lined paper, leaving plenty of space between columns, and plenty of space for additional rows.

1. Add rows for $x = 3$, 4, and 5. Compare the values of the three expressions now. Which one seems to be increasing most rapidly? Which one seems to be increasing least rapidly?

2. Add rows for $x = 6$, 7, and 8. Compare the values of the three expressions now. Which one seems to be increasing most rapidly? Which one seems to be increasing least rapidly?

3. Add rows for $x = 9$, 10, 11, and 12. Compare the values of the three expressions now. Which one seems to be increasing most rapidly? Which one seems to be increasing least rapidly?

4. Add a row for $x = 20$. Compare the values of the three expressions now. Which one seems to be increasing most rapidly? Which one seems to be increasing least rapidly?

5. What do you think will happen to each expression as the value of x gets even larger?

One way to measure the growth of these numbers is to look at how much they increase at each step. Display 2.28 is the same as Display 2.27 except that new numbers have been added to show the increases. For example, as x increases from 1 to 2, x^2 increases from 1 to 4, for a net increase of $4 - 1 = 3$. The new numbers are in **boldface type**.

x	$2 * x$		$2^\wedge x = 2^x$		$x^\wedge 2 = x^2$	
0	0		1		0	
		2		**1**		**1**
1	2		2		1	
		2		**2**		**3**
2	4		4		4	

Display 2.28

Add similar numbers showing increases to your own table. Discuss with a partner any patterns in the increases.

One way to analyze the increases is to see how they grow. Display 2.28 is the same as Display 2.27 except that new numbers have been added to show the increases in the increases. The new differences are in boldface type. The increases you calculated earlier are called **first differences**.

x	$2 * x$		$2\wedge x = 2^x$		$x\wedge 2 = x^2$	
0	0		1		0	
		2		1		1
1	2	0	2	1	1	2
		2		2		3
2	4		4		4	

Display 2.29

Add similar numbers showing increases in the first differences to your table.

1. What do you think these new differences are called?

2. Do you see any patterns in them?

3. What do you think would happen if you added many rows to your table and took third differences? Fourth differences? Twenty-ninth differences?

The numbers 2^x, where x is an integer, are called the **powers of two**. They are very important in the computer world. A digital computer is made up of many tiny switch like elements that are either on or off. These two possibilities are what makes 2 so important in computing. *Powers of ten* are also important because 10 is the base of our numeration system.

Find the first five powers of 10.

a

You can see that the powers of 10 increase even more rapidly than the powers of 2. You may have used a calculator that can raise numbers to a power to make your table of powers of 10. Here is another way to fill in the table. Each new power of 10 is just 10 times the last one. Thus, if we know that:

$$10\wedge 4 = 10^4 \text{ is } 10,000$$

then

$$10\wedge 5 = 10^5 = 10,000 * 10 = 100,000$$

Use this fact to find the next five powers of 10. You should be able to do this without using a calculator.

b

1. Do you find any patterns in the powers of 10? Use the pattern to find the next five powers of 10.

2. Use the pattern to find $10\wedge 100 = 10^{100}$.

3. How long did it take you to write down your answer?

4. If you had to use the number 10^100 = 10^{100} frequently, would you want to abbreviate it? How might you do this?

5. Use your calculator to find 10^100 = 10^{100}. Is this a better method than the one you used above?

Although it has its limits, your calculator is capable of dealing with very large numbers. Unfortunately, in our numeration system, very large numbers are represented by a great many digits. For example, consider the number 3^{100}. Using exponents gives us a very brief way of writing this number. If we write it in the usual way, we have

515377520732011331036461129765621272702107522001

This is too long for the calculator display. As a result, the calculator abbreviates it. You need to understand the calculator's abbreviation before it can be useful to you.

1. Find 3^{100} on your calculator. Does it agree with the result above?

2. Find the first five powers of 1000. Do you find any patterns in the results? Your calculator uses an abbreviation for some of the higher powers. You might be able to figure out how the abbreviation scheme works if you knew the right answer. How can you figure out the higher powers without your calculator?

Let's look at one of the powers of 2, such as 2^{40}. Our calculator displays this as
$$1.099511628\text{E}12$$

which represents the number
$$1.099511628 * 10^{12}$$

The "12" after the "E" is interpreted as an exponent of 10. Your calculator uses powers of 10 to abbreviate long numerals. These abbreviations are a form of **scientific notation**. Using this notation, large numbers are written as a small number times a power of 10. The small number always has exactly one (nonzero) digit to the left of the decimal point. The exponent of 10 must be an integer.

The numeral

$$1.099511624 * 10^{12}$$

is scientific notation. Its common name is

$$1,099,511,628,000$$

Is it *exactly* 2^{40}, or has it been rounded off? How do you know?

Here is some practice in identifying correct scientific notation:

Compute each of the following.

$$100 * 10^2 \qquad 10 * 10^3 \qquad 1 * 10^4 \qquad 0.1 * 10^5$$

Which of these expressions equals 10^4? Only one of these expressions is correct scientific notation for 10^4. Which one is it, and why?

Find four more ways to write 10^4 as a number times a power of 10.

You may wonder why this way of writing numbers is called "scientific." That is because science often involves very large numbers. For example, Display 2.30 gives the average distance of each of the planets from the Sun.

Planet	Distance from Sun, Miles		
Mercury	36,000,000	or	3.6×10^7
Venus	67,240,000	or	6.724×10^7
Earth	92,900,000	or	_____
Mars	141,710,000	or	1.417×10^8
Jupiter	483,880,000	or	_____
Saturn	_____	or	8.8714×10^8
Uranus	1,783,980,000	or	1.738398×10^9
Neptune	2,796,460,000	or	_____
Pluto	_____	or	3.666×10^9

Display 2.30

Your teacher will give you a copy of Display 2.30. Supply the missing numbers.

Sometimes we need to multiply large numbers together. For example:

$2000 * 30 = 60,000$ or $2 \times 10^3 * 3 \times 10^1 = 6 \times 10^4$

$20,000 * 30 = 600,000$ or $2 \times 10^4 * 3 \times 10^1 = 6 \times 10^5$

$2000 * 3000 = 6,000,000$ or $2 \times 10^3 * 3 \times 10^3 = 6 \times 10^6$

$200,000,000 * 300,000 = 60,000,000,000,000$ or

$2 \times 10^8 * 3 \times 10^5 = 6 \times 10^{13}$

Can you find a pattern in these examples? Does the following example fit the pattern? Explain.

a

$8000 * 3000 = 24,000,000$ or
$8 \times 10^3 * 3 \times 10^3 = 2.4 \times 10^7$

Scientific notation is also used to represent very small numbers. If we return to the powers of 10 (Display 2.31), we can see the following pattern. Each time the value of n increases by one, the value of 10^n is multiplied by 10.

n	$10^n = 10^{\wedge}n$
1	10
2	100
3	1000

Display 2.31

What happens to the value of 10^n each time n decreases by one?

b

If we continue the pattern backwards by one step from $10^1 = 10$ we get $10^0 = \dfrac{10}{10} = 1$. What happens if we go one step further beyond that?

According to the pattern you found, what should 10^{-1} be?

Copy Display 2.32. Use the pattern to finish the table. You should be able to do this without using a calculator. Note that we are counting down now and the order of rows is reversed.

n	$10^n = 10\wedge n$
2	100
1	10
0	1
−1	_____
−2	_____
−3	_____
−4	_____

Display 2.32

What do you think 1E-4 means?

Negative exponents also appear on calculators.

Use a calculator to find

$$\frac{1}{10} \qquad \frac{1}{100} \qquad \frac{1}{1000}$$

Continue the pattern until something interesting happens. Write a description of what happens and an explanation of what it means.

Negative exponents are often used with scientific notation for very small numbers. Display 2.33 gives some units of measure used for small things.

1 angstrom = 0.000000003937 or 3.937×10^{-9} inches
1 micron = 0.00003937 or 3.937×10^{-5} inches
1 mil= 0.001 or 1×10^{-3} inches
1 point = 0.013837 or 1.3837×10^{-2} inches

Display 2.33

Find out how each of these units of measure in Display 2.33 are used. Write a sentence about each. For example, the angstrom (abbreviated Å) is a unit used in measuring wavelengths of light. It is named after the Swedish scientist A. J. Ångström, who did research in this area in the 19th century.

X-rays, gamma rays, radio waves, and visible light are all part of the **electromagnetic spectrum.** Visible light has wavelengths from about 4000 Å (violet) to 6500 Å (red). Waves just longer than red are called **infrared.** Radio waves are even longer. Waves just shorter then violet are called **ultraviolet.** X-rays are even shorter.

Express the wavelengths of violet and red light (in inches) in scientific notation and in decimal fractions.

Problem Set: 2.6

1. How would you interpret 10E−4 if it showed up on your calculator? Is it the same as 10^{-4}?

2. Which of the following numbers are in scientific notation? For any that are not, explain why they are not.

$$1.34 * 10^{23} \qquad 0.34 * 10^{54}$$

$$1.34 * 1^{23} \qquad 10.34 * 10^{54}$$

$$1.34 * 11^{23} \qquad 7.34 * 10^{-54}$$

3. Use a calculator to find

$$\frac{1}{10^1} \qquad \frac{1}{10^2} \qquad \frac{1}{10^3}$$
$$\frac{1}{10^4} \qquad \frac{1}{10^5} \qquad \frac{1}{10^6}$$

   What is the pattern?

4. Express the following in scientific notation:

   (a) 150,000,000 km, the approximate distance from the Earth to the Sun.

   (b) 2,000,000,000,000,000,000,000,000,000,000 kg, the approximate mass of the Sun.

   (c) 0.0005 inches, the thickness of a piece of paper.

(d) 0.000000002 m, the width of a DNA molecule.

(e) 118,000,000,000 km, the diameter of the solar system.

(f) 8,500,000,000,000 km, the distance light travels in a year.

(g) 0.0000000000564 cm, the diameter of an electron.

(h) What is the advantage of scientific notation in these examples?

5. Express the following in scientific notation.

$$1,234,567,890 \quad 93,000,000 \quad 10^{34} \quad 10^{-34}$$

$$0.000000234 \quad \frac{1}{100,000} \quad \frac{2}{500,000} \quad 27$$

6. Choose the power of 10 that would give you the best estimate of each of the following quantities.

(a) population of the city of New York

(b) altitude in feet of Denver

(c) gross box office receipts of the movie *Spiderman 2*

(d) distance in miles from New York City to Los Angeles

(e) lifetime number of points scored by Michael Jordan

You may wish to use an atlas or almanac to check your answers.

7. If you multiply a power of 10 by a power of 10 you get another power of 10. Find a pattern in the examples in Display 2.34.

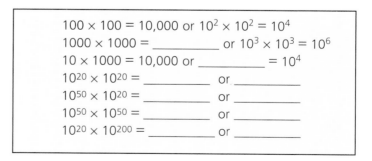

$$100 \times 100 = 10,000 \text{ or } 10^2 \times 10^2 = 10^4$$
$$1000 \times 1000 = \underline{\hspace{2cm}} \text{ or } 10^3 \times 10^3 = 10^6$$
$$10 \times 1000 = 10,000 \text{ or } \underline{\hspace{2cm}} = 10^4$$
$$10^{20} \times 10^{20} = \underline{\hspace{2cm}} \text{ or } \underline{\hspace{2cm}}$$
$$10^{50} \times 10^{20} = \underline{\hspace{2cm}} \text{ or } \underline{\hspace{2cm}}$$
$$10^{50} \times 10^{50} = \underline{\hspace{2cm}} \text{ or } \underline{\hspace{2cm}}$$
$$10^{20} \times 10^{200} = \underline{\hspace{2cm}} \text{ or } \underline{\hspace{2cm}}$$

Display 2.34

(a) Use the pattern to find the missing entries.

(b) Would the pattern you found work for powers of 2 as well as powers of 10? What about powers of other numbers?

(c) Can you express the pattern using algebra? If yes, do it. If not, explain why not.

(d) If the pattern works for fractional exponents, what would $10^{0.5} \times 10^{0.5}$ be? In that case, what would $10^{0.5}$ have to be?

8. Choose the best negative power of 10 to estimate the following lengths in meters.

(a) width of a needle

(b) height of an ant

(c) length of a TI-83 calculator

(d) thickness of a penny

(e) thickness of a TI-83 calculator

(f) length of a notebook

9. You may remember this children's verse. It is really a riddle and a trick question.

> As I was going to Saint Ives,
> I met a man with seven wives,
> Each wife had seven sacks,
> Each sack had seven cats,
> Each cat had seven kits:
> Kits, cats, sacks, and wives,
> How many were there going to Saint Ives?

(a) Answer the question in the riddle.

(b) How many were going the other way? Express your answer using exponents, then calculate a final answer.

PROJECT

If the distance from the Earth to the Sun is 1.5×10^8 km, how many steps would it take you to walk to the Sun?

2.7 Working With Exponents

Learning Outcomes

After studying this section, you will be able to:

State the Laws of Exponents

Demonstrate understanding the meaning of positive, negative and zero exponents

Simplify expressions involving positive and negative exponents

Apply the Laws of Exponents to express numbers in scientific notation

Apply the Laws of Exponents to algebraic expressions.

In this chapter you will see how algebra is used as a convenient way to express mathematical ideas in a shorter form than would be possible using words. You also learned how integer exponents are used in real life situations and how to use scientific notation to abbreviate very large and very small numbers.

Addition, subtraction, multiplication, and division have rules that govern how we deal with numbers when using these operations. The operation of raising a number to an exponent (exponentiation) also has its own rules. Recall that when we write something like 3^2 the number 3 is called the base, the 2 is the exponent and 3^2 is called a power. Note that if a number or a variable is written without any exponent, the exponent is understood to be 1.

To multiply terms in exponential form, we simply need to remember what the symbols mean:

$$3^4 \times 3^5 = 3 * 3 * 3 * 3 \times 3 * 3 * 3 * 3 * 3 = 3^9$$

$$5^3 \times 5 = 5 * 5 * 5 \times 5 = 5^4$$

For each of the following determine if the expression can be written as a single power. If yes, do so. If no, explain why not.

1. $2^3 \times 5^2$ 2. $2^2 \times 8^3$
3. $3^2 \times 27^2$ 4. $3^2 \times 6^2$
5. $2^2 \times 5^2$ 6. 6×3^2
7. 25×4^2 8. $5^2 \times 6^3$

Do you see a pattern? Explain.

Bases that are variables are treated in the same way as number bases. For example,

$$a^5 \times a^2 = a * a * a * a * a \times a * a = a$$

$$-3a^4 \times 2a^2 = -3 * a * a * a * a \times 2\, a * a = -6a^6$$

$$a^3b \times a^5b^2 = a * a * a * b \times a * a * a * a * a * b * b = a^8b^3$$

Facts to Know: To multiply two powers with the same base, add the exponents. $x^a * x^b = x^{a+b}$, where x, a, and b cannot all equal zero.

To multiply two powers with different bases and a common exponent, multiply the bases and raise the product to the common exponent.

$a^2 * b^2 = (a*b)^2 = (ab)^2$, where a and b cannot equal zero
$3^2 * 5^2 = (3 * 5)^2 = 15^2$

The statement $x^a * x^b = x^{a+b}$ is a *mathematical Law of Exponents*.

Write the following using $x^a * x^b = x^{a+b}$
(e.g., $3^3 * 3^4 = 3^7$)

1. $2^3 * 2^7$

6. $b^3 * b^7$

2. $4^3 * 4^2$

7. $5^4 * 5^5$

3. $x^4 * x^9$

8. $2a^2 * a^3$

4. $a^2 * a^4$

9. $-2x^3 * 3x$

5. $x^5 * x^3$

10. $a^2b^4 * a^3b$

Adding the exponents is a convenient way to multiply two powers with the same base. Because division is the inverse of multiplication what would we expect to do when dividing powers with the same base? Let's investigate by considering some examples:

$$\frac{2^{10}}{2^3} = \frac{2*2*2*2*2*2*2*2*2*2}{2*2*2} = 2*2*2*2*2*2*2 = 2^7$$

$$\frac{a^6}{a^3} = \frac{a*a*a*a*a*a}{a*a*a} = a*a*a = a^3$$

$$\frac{b^5}{b^3} = \frac{b*b*b*b*b}{b*b*b} = b*b = b^2$$

A Fact to Know: To divide two powers with the same base, subtract the exponent in the denominator from the exponent in the numerator and leave the base the same.

The statement $\dfrac{x^a}{x^b} = x^{a-b}$ where $x \neq 0$ is a *mathematical Law of Exponents*.

This idea makes sense because division is the opposite of multiplication and subtraction is the opposite of addition.

Write the following in simplified form (e.g., $\frac{5^7}{5^2} = 5^{7-2}, 5^5$).

a

1. $\frac{3^6}{3^3}$

2. $\frac{a^7}{a^4}$

3. $\frac{2^{10}}{2^5}$

4. $\frac{a^4b^7}{ab^2}$

5. $\frac{x^5}{x^5}$

6. $\frac{y^3}{y^7}$

7. $\frac{12x^3y^6}{3xy^2}$

8. $\frac{-25a^4b^4}{-5a^3b^2}$

9. $\frac{4^7}{4}$

10. $\frac{-24a^7b^4c^8}{-4abc^2}$

If you found the results for problems 5 and 6 troubling, take a few minutes and write down why you think these results may be difficult to understand. Explain what you think they might mean. We'll come back to them later.

Sometimes it is necessary to raise a power to an exponent. For example, $(a^4)^{12}$. We could do it by multiplying

$a^4 * a^4 * a^4 * a^4 * a^4 * a^4 * a^4 * a^4 * a^4 * a^4 * a^4 * a^4$ to get a^{48}.

It is easier to find the exponent 48 by recalling that multiplying 4×12 to get 48 is a shortcut to repeated addition.

$$(a^3)^4 = a^3 * a^3 * a^3 * a^3 = a^{12}$$

$$(2^5)^2 = 2^5 * 2^5 = 2^{10}$$

b

1. Use your calculator to evaluate $(3^4)^{10}$, remembering to key in 3^4 in parentheses first.

2. Use your calculator to evaluate 3^{40}. Do you get the same answer? Explain.

A Fact to Know: To raise a power to an exponent, multiply the exponents.

$(x^a)^b = x^{ab}$, where $x \neq 0$ when a or b equals zero.

In using this law, you must follow the convention about the *order of operations*. If there are any operations indicated in the exponent, we do those first. In x^{bc} we compute the exponent $b*c$ and then raise x to that exponent. As an example of the law, $(2^5)^3 = 2^{5*3} = 2^{15}$.

Keep in mind, too, that parentheses are used to clarify which quantities are being raised to a given exponent. For example, $3(x^2y^3)^2 = 3x^4y^6$ and $(3x^2y^3)^2 = 9x^4y^6$.

Recall that the symbol "−" means "find the opposite." So $-2^4 = -16$ because the expression says to raise 2 to the fourth power and then find the opposite. But, $(-2)^4 = 16$ because the expression $(-2)^4$ says to raise (-2) to the fourth power.

Use these ideas to simplify each of the following expressions:

a

1. $(x^3)^4$
2. $(a^6)^5$
3. -3^4
4. $(b^7)^4$
5. $(-2)^3$

6. $(a^2b^3)^4$
7. $(-3)^4$
8. $2(x^3y^2)^2$
9. $-3(xy^2)^3$
10. $(2ab^3)^3$

Here's an application of the use of the Laws of Exponents with scientific notation. Rupert Mudloch is a rich man who owns many TV stations and newspapers. He was unable to buy Lake Superior, one of the Great Lakes between Canada and the United States, so now he is creating a man-made lake that is square and 10^5 meters on a side. A competing media giant decides to build a square lake 3×10^5 meters on a side. What is the area of that lake?

To square 3×10^5 we need to calculate $(3 \times 10^5)^2 = 3^2 \times (10^5)^2 = 9 \times 10^{10}$ square meters. So, the competitor's lake will be 90,000,000,000 square meters in area!

Express each of the following in scientific notation:

b

(a) $(1.27 \times 10^{12})^5$

(b) $(2.56 \times 10^3)^2$

(c) $(2.03 \times 10^{-4})^3$

When we divided $\dfrac{x^5}{x^5}$ in an earlier problem using division with exponents the answer was x^0. We know that any nonzero number divided by itself equals 1, so $x^0 = 1$. Recall that in Section 2.6, we saw that we could preserve the pattern in our table of powers of 10 (Display 2.31) if we let $10^0 = 1$ and $10^1 = 10$. Both patterns and the division law for exponents confirm this result so we define $a^0 = 1$ where $a \neq 0$. But what about the value of a^0 if $a = 0$?

1. Calculate 1^1, 1^2, 1^3, and 1^4. Use algebra to express a general fact about 1^a. Write your conclusion with an exponent and also with a caret.

2. Calculate 0^1, 0^2, 0^3, and 0^4. Use algebra to express a general fact about 0^a. Write your conclusion with an exponent and also with a caret. According to this pattern, what should 0^0 be?

3. Calculate 1^0, 2^0, 3^0 and 4^0. According to this pattern, what should 0^0 be? There is no way to define 0^0 without breaking one of the patterns. For that reason, the value of 0^0 cannot be determined.

4. Use your calculator to calculate the value of 0^0. What happens? Explain.

Facts to Know:

- $1^a = 1$, for all values of a
- Any number (other than 0) raised to an exponent of 0 has a value of 1. $a^0 = 1$, $a \neq 0$
- 0^0 is undefined
- $0^1 = 0$

Recall you have seen negative exponents in powers of 10 when writing numbers in scientific notation. To establish what a negative exponent means in general, we can rewrite the original fraction as

$$\frac{x^6}{x^9} = \frac{1}{x^3}$$

It follows that

$$x^{-3} = \frac{1}{x^3}$$

A negative exponent represents the number of factors of a quantity that is being divided into 1 rather than the number of factors being multiplied by 1. We can think of powers with negative exponents as the reciprocals of powers with positive exponents. This result can now be used to establish a way to write the expression $\frac{1}{x^{-3}}$.

$$\frac{1}{x^{-3}} = \frac{1}{\frac{1}{x^3}} = 1 \times \frac{x^3}{1} = x^3$$

Facts to Know: Expressions involving negative exponents can be rewritten using positive exponents and vice versa.

$$x^{-a} = \frac{1}{x^a} \quad \text{and} \quad \frac{1}{x^{-a}} = x^a \text{ where } x \neq 0$$

We can also use these **Facts to Know** to reinforce what is meant by numbers written in scientific notation. For example, a number 5.3×10^{-9} is the same as $5.3 \times \frac{1}{10^9}$, and this is the same as $\frac{5.3}{10^9}$. If you do this division on your calculator the result will be 0.0000000053, and if this number is written in scientific notation it would be 5.3×10^{-9} —just as we expect!

For each of the following expressions in Display 2.35,

(a) Evaluate the expression with your calculator

(b) Write it with a positive exponent

(c) Evaluate the expression with a positive exponent with your calculator

(d) Verify that you get the same answer each time.

Expression	Calculator Value (a)	With Positive Exponent (b)	Calculator Value (c)
3^{-2}			
$\dfrac{1}{4^{-1}}$			
25^{-2}			
$\dfrac{1}{3^{-3}}$			
5^{-3}			
10^{-1}			
$\dfrac{1}{2^{-3}}$			
$\dfrac{1}{10^{-2}}$			
7^{-1}			
10^{-4}			

Display 2.35

REFLECT

In this chapter you have been introduced to the basic idea of algebra. We use letters to represent unspecified numbers (variables). This helps us to express general laws that are true for *any* numbers, such as the Associative and Commutative Laws of Addition and Multiplication, and the Distributive Law. We use algebra to express relationships between quantities in the world around us using formulas. In other situations, we have an algebraic equation that is true only for certain numbers, and we want to find those numbers. The process of finding them is called *solving the equation*.

In all of these situations, algebra provides an abbreviation for descriptions of the situation expressed in words. We learned about the arithmetic operation of exponentiation that abbreviates repeated multiplication and division. We used exponents in scientific notation, a way of abbreviating very large and very small numbers. We learned several Laws of Exponents.

Problem Set: 2.7

1. If you deposited $100 at a simple interest rate of 5% per annum, how much money would you have after 2 years? After 10 years? 20 years? 100 years? Answer these same questions for 5% compound interest per annum. Which type of interest would you prefer? Why? Does the advantage vary with time? Explain.

2. How much money would you have to deposit at a 3% simple interest rate to have $500 after 2 years? After 5 years? 10 years?

3. How much money would you have to deposit at a 3% compound interest rate per annum to have $500 after 2 years? After 5 years? 10 years?

4. The Commutative Law of Addition says that $a + b = b + a$ for all values of a and b. The Commutative Law of Multiplication says that $a * b = b * a$ for all values of a and b. If exponentiation were a commutative operation, we would have that $a{\char`\^}b = b{\char`\^}a$.

    (a) Rewrite $a{\char`\^}b = b{\char`\^}a$ using exponents instead of carets.

    (b) What do the laws have in common that makes us call them all "commutative"?

(c) Is exponentiation commutative? Why or why not?

5. List all the Laws of Exponents (see **Facts to Know**) you learned in this section. For each law, indicate whether it is true for all values of the variables. If not, explain the exceptions.

6. Can each of the following be expressed as a power of two? If Yes, do so. If No, explain why not.

(a) $2^3 * 2^5$

(b) $2^5 * 2^3$

(c) $2^{33} * 2^{85}$

(d) $2^{85} * 2^{33}$

(e) $2^{333} * 2^{857}$

(f) $2^{857} * 2^{333}$

(g) Write a law that summarizes what you found in these examples.

7. Can each of the following be expressed as a power of two? If Yes, do so. If No, explain why not.

(a) $2^{57} * 2^{85}$

(b) $\dfrac{2^{857}}{2^{333}}$

(c) $2^{857} * 2^{123}$

(d) $\dfrac{2^{857}}{2^{857}}$

(e) $2^{857} + 2^{857}$

(f) $2^{857} + 2^{333}$

(g) $\dfrac{2^{333}}{2^{857}}$

(h) $2^{333} + 2^{857}$

(i) $2^{333} + 2^{333}$

(j) 0

(k) 1

(l) 2

(m) 4

(n) 4^{333}

(o) $\dfrac{1}{2}$

(p) $\dfrac{1}{4}$

8. Can each of the following be expressed as a power of 10? If Yes, do so. If No, explain why not.

(a) $10^{333} * 10^{857}$

(b) $\dfrac{10^{857}}{10^{333}}$

(c) $10^{333} + 10^{857}$

(d) $10^{333} + 10^{333}$

(e) 0

(f) 1

(g) 10

(h) 1000

(i) 1000^{333}

(j) 0.001

(k) $\dfrac{1}{10}$

9. When sizes are very different, we usually use a factor to compare them. For example, in 2000 the population of the United States was about 281,421,906 while the population of Iran was about 65,619,636. It would give a better idea of the relative sizes of the two countries if we said that the U.S. had roughly four times the population of Iran rather than saying that the U.S. had 215,802,270 more people than Iran.

(a) In 2000, China had a population of about 1,261,832,482. How many times larger than the U.S. population was China's population at that time?

(b) For each of the following show how to use the Laws of Exponents to compare the larger distance to the smaller.

(i) The Earth is 1.5×10^{11} m from the Sun. The Moon is 3.8×10^8 m from the Earth. How many times farther away is the Sun than the Moon?

(ii) Saturn is 1.43×10^{12} m from the Sun. How many times further away from the Sun is Saturn than the Earth?

(iii) Pluto is 4.5×10^{12} m from the Sun. How many times further away from the Sun is Pluto than the Earth?

10. Use your calculator to compute 2^{-500}. Do you agree with the answer? Explain.

11. Zeno entered a local marathon. He ran the first half of the race in 2 hours, but that left him so tired he only ran half as far in the next two hours. In the two hours after that, he ran only half as far as that!

(a) How long is a marathon?

(b) How far did Zeno run after 2, 4, 6, 8 and 10 hours?

(c) How far did Zeno run after 20 hours? What was his average speed for the last two hours? How far is he from the finish line now?

(d) When do you think Zeno will finish the race?

Jonathan Rothstein
Artist and Mathematician

Architecture uses a mix of math and art. At first, it seems that these two just don't go together.

And yet, explains Jonathan Rothstein, a designer and an architect in California, "we use math at every stage in our process. Whether it's a one-family home or the awesome Guggenheim Museum in Spain, basic geometry rules the design."

As an art student at New York City's La Guardia High School, Jonathan discovered architecture. "I had some trouble reading in elementary school. So I became very visually oriented. I was always drawing, building and taking things apart. In 11th grade I took an architecture class and it really brought things together for me."

Jonathan learned that math is behind the creative design of every building and its site. In the land survey, for instance, every detail, such as each tree and slope, is calculated, then replicated in miniature. "Take the calculations necessary for slopes," he continues. "Slopes affect the drainage of, say, football fields and highways. You have to figure the slanting of house roofs, or of car garages, which are designed using a series of ramps. And, of course, the 1990 American Disabilities Act has brought ramping into many new areas of our lives.

"What I like more than anything," states Jonathan, "is working with people. I really like the process of creating something personal — together. As architects, we use math and art to produce buildings, public spaces and communities that are themselves works of art."

The Algebra of Straight Lines

CHAPTER 3

3.1 Coordinate Systems

Learning Outcomes

After studying this section, you will be able to:

Construct rectangular coordinate systems for the plane using various points of origin and unit lengths

Name points in the plane by using coordinate axes

Mark the locations of ordered pairs in the coordinate plane.

In Chapter 2 you learned how to represent numbers, operations, and other ideas with letters and symbols. In this chapter, we add a special kind of drawing to your mathematical tool kit. You'll see how something called a *coordinate system* combines algebra with the picture power of geometry. You probably have seen coordinate systems before, even if you never studied them in school. In fact, you use one every time you read a map. We'll start there and gradually sharpen the simple, basic idea into a powerful mathematical tool.

On the U.S. map in Display 3.1, Denver, Colorado, is in 4C and Atlanta, Georgia, is in 8D.

1. Where is Portland, Oregon?

2. Where is Washington, D. C.?

3. What city is in 6B?

4. Describe how this code works. What do the numbers represent? What do the letters represent?

5. What is the code symbol for your home town?

6. Is there exactly one number letter code symbol for every city named on the map? For every city in the United States? Which cities might be a problem? (*Hint:* Look at Houston, Texas.)

7. If someone gives you a number letter code symbol, can you pick out the city that person has in mind? Try 9B.

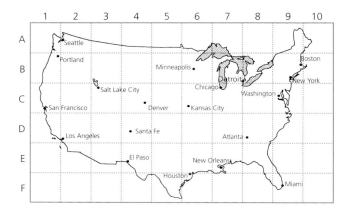

Display 3.1

About Words

In English, the prefix *co* signals a person or thing that works along with another person or thing of the same kind, as in *co-captain* and *coauthor*. The *-ordinate* part means something related to order. (What is an *ordinal* number?)

The number letter code used in map reading is an example of a **coordinate system**, a way of using two or more numbers or letters to locate a position in space. Each of the numbers or letters used to identify the position is called a **coordinate** of that position.

The coordinate system in Display 3.1 is crude. It divides the map of the United States into 60 squares. Each number letter pair identifies a square. The area covered by a square is very big. It contains places that are many miles apart. For example, the square 9B contains New York City, Boston and most of Massachusetts, all of Connecticut and Rhode Island, and parts of six other states. Can you name them?

Two small changes in this coordinate system will allow us to locate places more exactly. First, instead of using a letter and a number, we'll use two numbers. Second, we'll think of each pair of coordinates as naming the lower left corner of a square. That is, instead of having each coordinate name a strip of the map, we'll have it name one of the dividing lines. We'll start the numbering at the lower left corner, using 0 as the first number for both the horizontal and vertical scales. See Display 3.2.

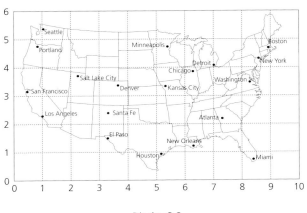

Display 3.2

Changing from number letter coordinates to number number coordinates causes a small problem. In Display 3.1, find the place with coordinates C and 6 (or 6 and C, if you prefer). In Display 3.2, find the place with coordinates 3 and 2 (or 2 and 3, if you prefer). What's the problem? How would you solve it?

When both coordinates are numbers, we have to know which one refers to the scale along the bottom of the map and which one refers to the scale up the side. For instance, 3 on the bottom and 2 on the side of Display 3.2 denotes the *lower left corner* of the square containing Santa Fe, New Mexico. But 2 on the bottom and 3 on the side denotes the square containing Salt Lake City. Which state is Salt Lake City in?

The usual way to fix this difficulty is to rely on the order in which each pair of coordinates is stated. We shall agree that the first number is used for the bottom (horizontal) scale and the second number is for the side (vertical) scale. A pair in which the order matters is called an **ordered pair**. To write two numbers as an ordered pair, separate them by a comma and put them in parentheses. The ordered pair with first number 3 and second number 2 is written (3, 2). It is *not* the same as (2, 3).

These questions refer to the map in Display 3.2.

1. Which square has (3, 1) as its lower left corner? What city is in it?

2. Which square has (1, 3) as its lower left corner? Which state makes up most of this square?

3. What are the coordinates of the lower left corner of the square containing the southern tip of Texas?

4. If you reverse the coordinates of your answer to question 3, which square do you get? Can you name a city in it?

5. What is the lower left corner of the square containing Yellowstone National Park?

6. If you reverse the coordinates of your answer to question 5, which square do you get? Can you name four states that are partly in it? If yes, do it.

7. What are the coordinates of the square containing your school?

8. If you reverse the coordinates of your answer to question 7, which square do you get? Can you name some place in it? If yes, do it.

Look again at New York City and Boston. They're both in the square with lower left corner (8, 4), but they are more than 200 miles apart. They ought to have different coordinate pairs. One way to get different pairs is to divide the (8, 4) square into smaller squares and use their lower left corners. Display 3.3 shows how this looks when the unit lengths are divided into tenths. Now Boston is in the square (8.8, 4.6) and New York City is (mostly) in (8.5, 4.2)

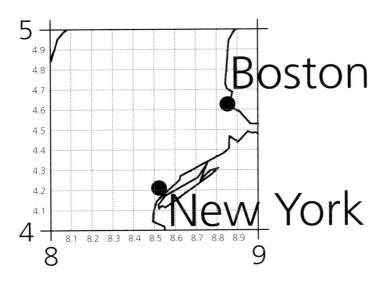

Display 3.3

There is a more precise way to label each point with its own coordinate pair.

1. Draw lines directly from the point to each of the two scales on the edge of the picture. By "directly" we mean that your lines meet the edges at right angles; they are *perpendicular* to the edges.

2. On each scale, find the number where the perpendicular line meets it. That is, find the distance from that intersection point to 0.

3. Use these two distances as the coordinate of the point.

About Words

The word *dimension* comes from the Latin word *dimensus*, which means measured separately. A plane is called two dimensional because the location of each point requires two separate measurements.

For example, look at the location of Albany, New York, in Display 3.4. To find its coordinate pair, we draw from its spot a vertical line to the bottom scale and a horizontal line to the left side scale. The vertical line meets the bottom scale at about 8.43; the horizontal line meets the left scale at about 4.57. Thus, the ordered pair (8.43, 4.57) is—approximately—the coordinate pair for Albany.

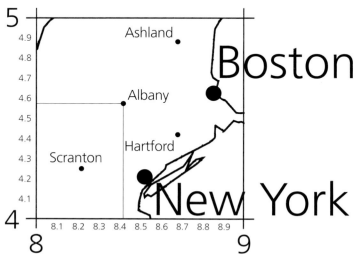

Display 3.4

We say "approximately" because we just estimated, to two decimal places, where the horizontal and vertical lines meet the scale lines. These approximations can be made as precise as eyesight and drawing tools will allow. In theory, there is an exact pair of numbers for each point on the map. In practice, a reasonable approximation is usually good enough.

These questions refer to Display 3.4.

1. What is the coordinate pair of Scranton, Pennsylvania?

2. What is the coordinate pair of Hartford, Connecticut?

3. Where is the point (8.68, 4.88)? Find it on the map. What state do you think it's in?

4. Where is the point (8.85, 4.12)? Find it on the map. What can you say about this location?

Look back at the map in Display 3.2. Using the same scale lines, how could it be extended westward to include Hawaii and Alaska?

This method of labeling points by pairs of numbers can be used for any planar (flat) region. The Earth is not flat, of course. Maps of regions on it are adjusted so that the region can be treated as if it were flat. All that is needed is a pair of scale lines to act like the edges of the map. If such lines are not already given to you, you can make them yourself.

- Draw a horizontal line and a vertical line on the plane of a piece of paper, anywhere you like. We'll call any flat surface a *plane* from now on. Each of these lines is called a **coordinate axis** (or just an **axis**).

- Pick a convenient length for your **unit length** — the distance on each axis between 0 and 1, between 1 and 2, etc. You could use a different unit length for each axis. The process would work just as well, but the shapes would look different. Sometimes using different unit lengths for each coordinate is useful. For now, however, use the same unit length for both axes, unless you are told otherwise.

- Label the crossing point 0 on each axis. This point is called the origin of the coordinate system.

- Using your unit length, mark off the integer points on the axes. On the horizontal axis, put the positive numbers to the right of 0 and put the negative numbers to the left of 0. On the vertical axis, put the positive numbers above 0 and put the negative numbers below 0.

- To find the number that corresponds to any other point on an axis, use your unit length to measure the distance between that point and the origin.

That's all there is to it. Now you have a system that labels every point on the plane by exactly one ordered pair of real numbers. This means that you should be able to

(a) find the coordinate numbers for any point someone picks out, and

(b) find the point corresponding to any ordered pair of numbers.

About Words

The plural of *axis* is *axes*.

This system for giving each point a two number address is an easy way to link geometry with algebra. It lets us use tools from each of these areas to work on problems in the other. This powerful connection was first used effectively in 1637 by a French scholar named René Descartes. For this reason, it is sometimes called a **Cartesian coordinate system**. Because the axes cross at right angles, it is also called a **rectangular coordinate system**.

Problem Set: 3.1

You will need a ruler for some of these problems.

1. Display 3.5 is part of a plane with some points labeled by letters. The bottom and left sides of the frame are marked off as coordinate axes. Use these axes to write the coordinate pair of each labeled point.

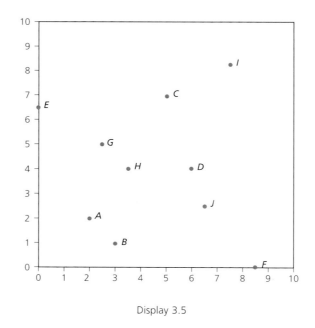

Display 3.5

2. Display 3.6 is part of a plane with some points labeled by letters. Your teacher will give you a copy.

   (a) Choose one of the labeled points to be the origin of your coordinate system and draw two axis lines through that point. How might you make sure that your lines cross at right angles?

(b) Pick a convenient unit length and mark off the integer points on each coordinate axis.

(c) Using your coordinate system, write down the coordinate of each point that is marked with a letter. Estimate the numbers to one decimal place.

(d) Mark the location of each of these points:

P: (2, 1)   Q: (0.5, 0.75)   R: (−1, 1)   S: (−1, −0.5)

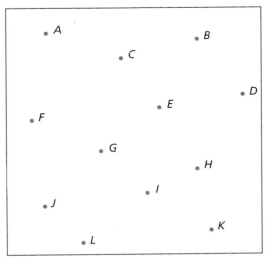

Display 3.6

3. Do problem 2 again, using the same unit length but choosing a different point for the origin.

4. Do problem 2 again, using the same origin but picking a different unit length.

5. Draw a coordinate system, choose a unit length, and mark the locations of each of these points

(2, 4)   (−3, 1)   (−3, −3)   (2, −4)

(0.7, 1.5)   (−3.75, 4.1)   (2.6,−1)   (−5.4, −3.25)

6. Suppose that you have a coordinate system with 1 centimeter as its unit length. Three points on the plane—A, B, and C—have these coordinates in that system

   A: (37, 50)   B: (100, 360)   C: (-256, 708)

   (a) If the unit length is changed to millimeters, what are the new coordinates of A, B, and C?

   (b) If the unit length is changed to meters, what are the new coordinates of A, B, and C?

7. Draw a coordinate system that has 1 cm as the unit length on each axis. Mark the locations of these points

   A: (1, 1)   B: (4, 1)   C: (4, 4)   D: (1, 4)

   Then draw the straight line segments AB, BC, CD, and DA.

   (a) Is the figure ABCD a rectangle? Why or why not?

   (b) Is the figure ABCD a square? Why or why not?

   Now draw a coordinate system that has 1 inch as the unit length on the x-axis and 1 cm as the unit length on the y-axis. Mark the locations of points A, B, C, and D in this coordinate system. Then draw the straight line segments AB, BC, CD, and DA.

   (c) Is the figure ABCD a rectangle? Why or why not?

   (d) Is the figure ABCD a square? Why or why not?

3.2 Plotting Sets of Points

People have settled on a custom to avoid having to say horizontal or vertical each time they want to talk about a coordinate axis. The horizontal axis is called the **x-axis**; the vertical axis is called the **y-axis**. We have already agreed that the ordered pair for a point should list the horizontal coordinate first, so we use (x, y) to represent a typical, but unknown, point in the plane. Often the first coordinate of a point is called its **x-value** and the second coordinate is called its **y-value**.

To **plot** a point means to use its coordinates to find its location, like this

1. Locate the first coordinate on the x-axis and the second coordinate on the y-axis.

2. Draw lines perpendicular to the axes at these coordinate points.

3. Find where these lines cross; that's where the point is.

Using 1 cm as the unit of measure, draw an x-axis from ⁻6 to 10 and a y-axis from ⁻2 to 10. Mark each axis at every cm point.

1. Plot this set of points: {(2, 0), (2, 5), (1.5, 5), (5, 8.5), (5.5, 8), (5.5, 8.5), (6.5, 8.5), (6.5, 7), (8.5, 5), (8, 5), (8, 0)}.

2. Draw straight lines to connect the points you plotted, taking them in the order of their listing in 1. What outline do you see?

3. Write down the coordinates of four points that, when connected, outline a door in an appropriate place in the picture. Then plot and connect them.

4. Write down the coordinates of four points that, when connected, outline a window in an appropriate place in the picture. Then plot and connect them.

5. The owner wants to put a TV antenna on the roof halfway up the left side. Find the coordinates of the point where that would appear on the outline. Explain how you found these coordinates.

Learning Outcomes

After studying this section, you will be able to:

Plot sets of points in the coordinate plane

Use coordinates to describe lines and segments that are parallel to the x-axis or the y-axis

State inequalities in algebraic notation and interpret them as regions of the coordinate plane

Use set-builder notation to describe sets of points.

6. There are 10 line segments in the original outline. For each one, find the coordinate of a point on it that is not one of the endpoints.

7. Write down the coordinates of some points that, when connected, outline a doghouse on the left side of the *y*-axis. Then plot and connect them.

8. What other features can you add to this picture by plotting points and connecting them with straight lines? Try some.

Pictures in the plane are sets of points. A coordinate system lets us describe points as ordered pairs of numbers. This means that we can describe a picture as a set of ordered pairs. This is how a TV or computer screen "thinks" about pictures. Now, that's easy enough to do when the picture is just a few isolated points; we just list the ordered pair for each point. But what if the picture is a figure with lines or segments or curves in it? Even a simple line segment contains *lots* of points—infinitely many, in fact—and we can't possibly list all of them!

Can you think of a way to describe *all* the points on a particular line without drawing a picture? To test whether or not your description is good enough, ask yourself this: If somebody else gives me the coordinates of a point, will my description always tell me whether or not that point is on the line?

To specify which points are on a line or segment and which are not, we have to find a test condition. That is, we need a statement about a typical ordered pair (x, y) that is

- true whenever the coordinates of a point in the figure are put in for *x* and *y*, and

- false whenever the coordinates of a point not in the figure are put in for *x* and *y*.

Both of these conditions must hold in order for us to have a good description of the set of points that makes up the figure.

Before we tell you any more, try this example. Find a condition that describes all the points of the line segment in Display 3.7, and no others. Start by answering these questions:

Display 3.7

1. What are the coordinates of 10 different points on this line segment?

2. If (x, y) is any point on the segment, what *must* its y-coordinate be?

3. If (x, y) is any point on the segment, what is the smallest value that its x-coordinate can have?

4. If (x, y) is any point on the segment, what is the largest value that its x-coordinate can have?

Now, what statement about x and y will be true for every point (x, y) that is on this line segment and will be false for every other point?

How did you do? Did you find a statement about the coordinates that applies to every point on that segment and to no others? Here's one.

$y = 2$ and x is a number between 1 and 5.

Remember that "and" means that *both* parts of the statement must be true.)

Here's a chance to apply what you just saw. Find a statement about coordinates that is true for every point of the line segment in Display 3.8 and is false for every other point.

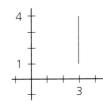

Display 3.8

The problem with a statement such as "x is a number between 1 and 5" is that "between" doesn't say whether or not x can actually be 1 or 5. There are some useful symbols that clarify such situations without a lot of words. To see how they work, we start with the numberline picture of "x is between 1 and 5" in Display 3.9.

x is between 1 and 5

Display 3.9

About Symbols

The symbol < is a lot like the musical sign for *crescendo*,

which means, go from softer to louder.

Now, the symbol < means "is less than," so we can write "1 is less than 5" as 1 < 5. If we want x to be *strictly* between 1 and 5—that is, if we don't want to let x equal either end number—we write

$$1 < x < 5$$

This says

1 is less than x, and x is less than 5.

On the other hand, if we want to allow 1 and 5 as possible values for x, we should say

1 is less than or equal to x, and x is less than or equal to 5.

The symbol for "less than or equal to" is ≤, so we can write

$$1 \leq x \leq 5$$

Use the symbols < and ≤ to describe the segment of Display 3.8 in two ways:

1. **with the endpoints included, and**

2. **without the endpoints.**

About Symbols

The symbol > is a lot like the musical symbol for *decrescendo*,

which means, go from louder to softer.

The opposite to < is the symbol >, which means "is greater than." Again, putting a bar under it allows equality. Thus, we could write $5 \geq x \geq 1$ or $5 > x > 1$. However, most people prefer the *less than* forms because they match the usual left-to-right ordering of the number line and the order in which we read English.

Describe the following segments of the number line in words. If you use the word *between*, be sure to say which endpoints are included.

1. $-1 \leq x \leq 8$
2. $3 \geq x > 0$
3. $0 < x \leq 35$
4. $-4 \leq x < 3$
5. $-100 < x < -10$

An algebraic statement containing any of the symbols $<$, $\leq$, $>$, or $\geq$ is called an **inequality**. It says that two numbers or expressions representing numbers are *not* (or may not be) equal by telling you which *is* (or may be) the larger one. Even though $1 < x < 5$ is actually two inequalities combined, such a statement is often called an "inequality," too.

About Words

The prefix *in-* sometimes means not, as in *invisible, incorrect, inequity,* etc.

None of the following inequalities is correctly stated. What is wrong with each one?

1. $5 < x < 3$
2. $-3 \leq x \leq -5$
3. $5 > x < 10$
4. $5 \leq x \geq 10$

In the plane, the points for which one coordinate is a particular value (and the other coordinate can be any number) is a line parallel to an axis. Display 3.10 shows the line $y = 2$ (which is parallel to the x-axis). The points for which one coordinate must be between two numbers (and the other coordinate can be any number) is a strip parallel to an axis. The shaded part of Display 3.10 shows the strip for $1 \leq x \leq 5$. Requiring *both* of these conditions means that we get only the points where the line and the strip overlap: the line segment of Display 3.7. In other words, this line segment is described by

$$y = 2 \text{ and } 1 \leq x \leq 5$$

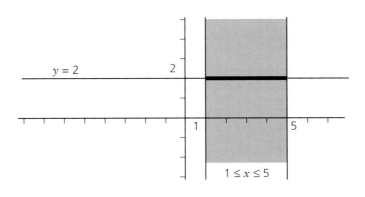

$y = 2$

$1 \leq x \leq 5$

Display 3.10

What strip and what line overlap to form the segment of Display 3.8? Make a sketch. Label the line and the strip in your sketch with algebraic conditions, as in Display 3.10.

1. What condition on the coordinates describes all the points of the shaded rectangular region shown in Display 3.11 and no other points?

2. What if you don't want to include the points of the boundary rectangle?

3. What if you want *only* the points of the boundary rectangle?

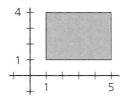

Display 3.11

The line segment of Display 3.7 is the set of all points (x, y) satisfying the conditions $y = 2$ and $1 \leq x \leq 5$. The line segment of Display 3.8 is the set of all points (x, y) satisfying two similar conditions. To avoid writing out phrases such as "the set of all points (x, y) satisfying" again and again, people have agreed on a way to abbreviate them with symbols. This shorthand form is called **set-builder notation** because it describes how to build a set from the conditions that its elements must satisfy.

The notation looks like this

$$\{[\text{element form}] \mid \dots [\text{conditions}] \dots \}$$

• The braces, { }, tell you that a set is being described.

• The dividing line, $\mid$, stands for "such that."

• The symbols in front of the dividing line represent a typical element of the set. For points in the plane, the form would be an ordered pair, such as (x, y).

• The conditions can be equations, inequalities, or any other statements that the elements must satisfy in order to be in the set.

For example, the segment of Display 3.7 is

$$\{(x, y) \mid y = 2 \text{ and } 1 \leq x \leq 5\}$$

In this case, we can abbreviate even further, if we want. Since y must be 2, we can write

$$\{(x, 2) \mid 1 \leq x \leq 5\}$$

1. Describe the line segment of Display 3.8 in set-builder notation in two different ways.

2. Translate into words

$$\{(x, y) \mid x = 4 \text{ and } -1 \leq y \leq 3\}$$

then draw this set on a coordinate plane.

3. Draw the set $\{(x, 3) \mid -5 \leq x \leq -2 \text{ or } 2 \leq x \leq 5\}$ on a coordinate plane.

Problem Set: 3.2

You will need to set up coordinate systems for this problem set. Graph paper will make this easier to do, but you can use plain paper if you prefer.

1. Draw the following sets of points on a single coordinate system with axes stretching from -10 to 10. Label each part on your drawing. Also rewrite each description in set-builder notation.

   (a) The set of all points (x, y) such that $x = 4$ and y is between 2 and 8, inclusive.

(b) The set of all points (x, y) such that $y = 4$ and x is between 2 and 8, inclusive.

(c) The set of all points (x, y) such that $y = -6$ and $-3 \leq x \leq 3$.

(d) The set of all points (x, y) such that $x = -3$ and $-9 \leq y \leq -6$.

(e) The set of all points (x, y) such that $x = -2$ and $y \geq -3$.

(f) The set of all points (x, y) such that $y = -2$ and $x \leq -3$.

(g) The set of all points (x, y) such that $y = -9$ and $x \geq -7$.

2. Translate each set-builder description into words.

(a) $\{(x, 1) \mid -2 \leq x \leq 6\}$
(b) $\{(3.5, y) \mid y \geq 1.5\}$
(c) $\{(x, 3.5) \mid x < 1.5\}$
(d) $\{(x, y) \mid 2 < y < 3\}$

3. (a) $\{(x, 7) \mid -5 \leq x \leq 4\}$ is a line segment. State whether each of the following points is on or off it.

(-5, 4) (0, 7) (7, 0) (3.25, 7) (-2.13, 7)
(1.86, 3.9)

Write the coordinates of three other points that are on this segment and three other points that are off it.

(b) $\{(6, y) \mid -1.5 \leq y \leq 2.7\}$ is a line segment. State whether each of the following points is on or off it.

(0, 0) (0, 6) (6, 0) (-1.5, 6) (6, 2.7) (6, 3)

Write the coordinates of three other points that are on this segment and three other points that are off it.

4. Draw a coordinate system with each axis stretching from -7 to 7. Then draw in the following four strips. Shade the strips to show clearly where they overlap.

(a) $\{(x, y) \mid 2 \leq x \leq 6\}$
(b) $\{(x, y) \mid 3 \leq y \leq 5\}$
(c) $\{(x, y) \mid -5 \leq x \leq -4\}$
(d) $\{(x, y) \mid -7 \leq y \leq -4\}$

5. Write a set-builder description for line parts *a–h* of Display 3.12.

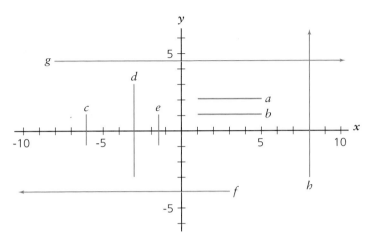

Display 3.12

6. A TV screen works like a coordinate plane. Each tiny dot that lights up has its own two-coordinate address. For these questions, assume that the TV design engineers are working on a split-screen design to show four different pictures at once. They have set up the coordinate axes with the origin right in the center of the screen. Display 3.13 shows their setup, with the four picture locations numbered I, II, III, and IV. The coordinate axes are not in any of the pictures.

   (a) What conditions on x and y put a point (x, y) in picture I?

   (b) Specify the conditions for (x, y) to be in each of the other three pictures.

   (c) The engineers want to allow for a "Newsflash!" box right in the center of the screen, 8 units wide and 6 units high. The unit length varies with the size of the screen, but you can think of inches or centimeters, if you like. What conditions on x and y put a point (x, y) inside that box?

   (d) The whole screen is 20 units wide and 16 units high. An announcement box 6 units wide and 3 units high is to be put in the lower left corner of the screen. What conditions identify the points inside that box?

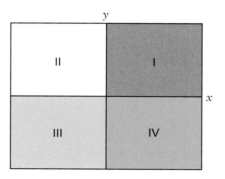

Display 3.13

7. Three corners of a rectangle are at $(-2, 3)$, $(5, 3)$, and $(-2, -1)$.

   (a) Find the coordinates of the fourth corner. Then set up a coordinate system and plot all four corner points.

   (b) Write a set-builder description of each side of the rectangle.

   (c) Write a set-builder description of the rectangular region.

3.3 Straightness

What does it mean to say that a line is straight? Describe "straightness" in as many different ways as you can.

So far, we have algebraic descriptions of lines, segments, and rays only if they are parallel to one of the coordinate axes. But most lines are not like that; what about them? How can we use algebra to describe these other lines, segments, and rays?

An ant is at (0, 0), the origin of a coordinate plane. It wants to walk to (10, 5) by the shortest path, which is along a straight line. Attached to one ankle it has an ant-sized electronic gadget that tells it the exact coordinates of each point it steps on. How can it use this gadget to stay on the straight path from (0, 0) to (10, 5)?

1. On a piece of graph paper, draw a pair of coordinate axes. Mark the point (10, 5).

2. Use a ruler to draw a straight path from (0, 0) to (10, 5).

3. Draw three other paths from (0, 0) to (10, 5), any way you want.

4. Mark four points on each path you drew. Use the coordinates of these points to fill in the four small tables in Display 3.14. As an example, one point of the straight path, (2, 1), has been picked for you and filled in.

5. Do any of the four tables have an obvious pattern? If so, describe the pattern(s).

6. What advice would you give to the ant?

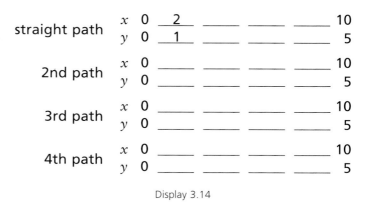

straight path	x	0	2	___	___	___	10
	y	0	1	___	___	___	5
2nd path	x	0	___	___	___	___	10
	y	0	___	___	___	___	5
3rd path	x	0	___	___	___	___	10
	y	0	___	___	___	___	5
4th path	x	0	___	___	___	___	10
	y	0	___	___	___	___	5

Display 3.14

Here's a situation from everyday life that provides an important clue to the connection between straight lines and the coordinates of points on them.

The door to the factory outlet store at the Fuzzy Friends Toy Co. is reached by climbing the stairs to a loading platform that is 4 feet above the level of their parking lot. (Display 3.15). To give delivery carts easy access to the main level, the company has decided to build a ramp down the side of its building to the opposite corner, 32 feet away from the edge of the platform. They need to support the ramp at 4 foot intervals. How high should they make each of the 7 supports?

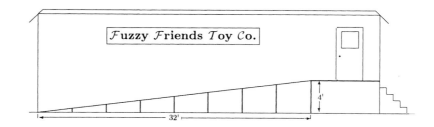

Display 3.15

Thinking Tip

Make a picture.
A picture or diagram of a situation often leads to interesting ideas and helpful ways of thinking about problems.

How would *you* handle this problem? Where would you start? Can you draw a simpler diagram? What are the important features of the picture in Display 3.15, and what can be left out?

Display 3.16 shows how you can use a coordinate system to restate the problem in a convenient way. If you think of the level of the parking lot as the *x*-axis and the lowest point of the ramp as the origin, then the problem becomes finding the missing *y*-coordinates of each of the 7 points marked along the ramp.

- What are they?

- How did you find them?

- Are you sure that the values you found will make the ramp straight? Why or why not?

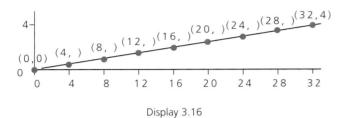

Display 3.16

The following questions focus on a pattern that relates the coordinates of points on a straight line to each other. These questions all refer to one particular line, but the same kind of pattern works for many different lines. See if you can figure out the general pattern as we go along.

Draw a coordinate system on a piece of paper, and then draw the straight line through (0, 0) and (1, 2).

1. Can you find the coordinates of another point on this line? What are they? Are you sure that it's really on the line, not just close?

2. Can you find the *y*-value of a point (2, *y*) on this line? If so, what is it? If not, why can't it be done?

3. Find the *y*-value of a point (3, *y*) on this line. Do you see a pattern here? What is it? If you see a pattern, use it to find the *y*-value of a point (10, *y*) on this line. Then use it to find the *y*-value of a point (-20, *y*) on this line.

4. Can you find the *y*-value of a point (3, *y*) that is very close to this line but not on it?

5. Can you find the *x*-value of a point (*x*, 3) on this line? If so, what is it? If not, why can't it be done?

6. Find the *x*-value of a point (*x*, 10) on this line. Do you see a pattern here? What is it? If you see a pattern, use it to find the *x*-value of a point (*x*, 50) on this line. Then use it to find the *x*-value of a point (*x*, -10) on this line.

Thinking Tip

Look for a pattern.
If you find one, try to figure out why it works. Patterns rarely occur by accident; they usually signal that something interesting is going on.

7. Try to describe in words a condition that will test the coordinates of a point to see whether or not it is on the line.

Can you generalize the pattern of the previous questions to another line? Try it. Redo questions 1–7 above, but *start with two different points.* Does your pattern still work?

• If it does, try to write it down in a way that applies to *any* straight line.

• If it doesn't, try to figure out what went wrong when you changed the coordinates of the two points. What happens if you leave one of the points the same but change the other?

To see how the coordinate pattern for straightness works, let's go back to the ramp at the Fuzzy Friends Toy Co. If the ramp has to be 4 feet high at the end of 32 feet, then how high should it be at 16 feet, the halfway point? Shouldn't it be 2 feet high there, exactly half as high as it needs to be at the end? Sure! If it didn't get that high in the first half, then the second half would have to be steeper. And if it were higher than 2 feet at the halfway point, then the second half would not be as steep as the first half. Either way, the ramp wouldn't be straight (Display 3.17).

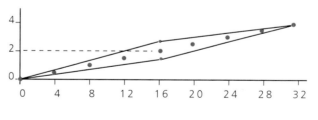

Display 3.17

Now generalize this reasoning,

• At 8 feet ($\frac{1}{4}$ of the way), the ramp should be 1 foot high.

• At 4 feet ($\frac{1}{8}$ of the way), the ramp should be $\frac{1}{2}$ foot high.

• . . . ?

At 1 foot ($\frac{1}{32}$ of the way), how high should the ramp be? How did you get your answer?

a

The general principle at work here is this: In order for the ramp to be straight, it must rise exactly the same amount for each unit of its horizontal length. To find how much that amount is, just divide the total rise by the number of units of horizontal length.

For the Fuzzy Friends ramp, we divide 4 feet by 32 feet to see that the ramp must rise $\frac{1}{8}$ of a foot for every foot of its horizontal length.

Now that we know the amount of rise per (horizontal) foot, we can find the height of the ramp anywhere along the side of the building just by multiplying the horizontal distance from the corner (measured in feet) by this amount. For instance, at 24 feet the ramp must be 3 feet high because $\frac{1}{8} \times 24 = 3$.

How can you use the rise per foot to tell how much the ramp rises between any two ground points? In particular, start at a point *A* on the ramp that is 10 feet from the left corner of the building (measured horizontally). Now move to a point *B* that is 22 feet from the left corner (measured horizontally). How much has the ramp risen in that span (Display 3.18)?

b

This example illustrates the key to describing straightness by coordinates. The main idea is

Between any two points on a straight line, the rise per horizontal unit must be the same as the rate of rise between any other two points on that line.

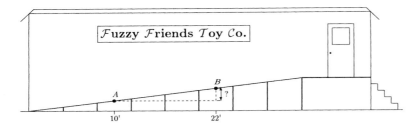

Display 3.18

In other words, the ratio of vertical change to horizontal change must be a *constant* (a fixed number). This is one of the most important mathematical ideas that you will see this year. It has a name that may remind you of the ramp at Fuzzy Friends Toy Co. every time you see it!

> **A Word to Know:** The **slope** of a straight line is the ratio
>
> $$\frac{\text{change in } y}{\text{change in } x}$$
>
> between any two points on the line.

We can compute the slope of the ramp in Display 3.18 by using the points A and B, which have coordinates (10, 1.25) and (22, 2.75)

$$\text{change in } y = 2.75 - 1.25 = 1.5$$

$$\text{change in } x = 22 - 10 = 12$$

$$\frac{\text{change in } y}{\text{change in } x} = \frac{1.5}{12} = \frac{1}{8}$$

If you use your calculator to divide 1.5 by 12, you get 0.125.

The definition of slope is sometimes abbreviated as "rise over run." The rise is the vertical change, and the run is the horizontal change. Think of the ramp example. If the line slopes downward as it goes from left to right and if the two points are taken in that order, then the rise is a negative number and the run is positive. Thus, the slope of such a line is negative.

Thinking Tip

Find or make examples. Often a general question is much easier to answer after you have explored some particular instances of it.

For example, let us find the slope of the line containing the points (3, 4) and (1, 5), shown in Display 3.19.

$$\text{change in } y = 4 - 5 = -1$$

$$\text{change in } x = 3 - 1 = 2$$

$$\frac{\text{change in } y}{\text{change in } x} = \frac{-1}{2} = -\frac{1}{2}$$

Display 3.19

Do you get the same result if you use the points in the opposite order? Try it by copying and completing these steps.

change in y = _____ = _____

change in x = _____ = _____

$$\frac{\text{change in } y}{\text{change in } x} = \text{_____} = \text{_____}$$

Did you get the same number for the slope? You should have. If you didn't, can you explain what went wrong?

1. When we compute the slope of a line from two points on it, does the order in which we use the points *ever* affect the result? Why or why not?

2. Does *every* straight line have a slope? Justify your answer.

The idea of slope makes it very easy to describe straight lines through (0, 0). Let's look at an example. Display 3.20 shows the straight line through (3, 2) and (0, 0). Its slope is

$$\frac{2-0}{3-0}$$

That is, its slope is just the ratio formed by the y-value of (3, 2) divided by its x-value. But the slope of a line is a constant. This means that the ratio $\frac{y}{x}$ for *any* point (x, y) on this line must be $\frac{2}{3}$

because we could use the coordinates of that point along with (0, 0) to compute the slope.

$$\frac{y - 0}{x - 0} = \frac{y}{x}$$

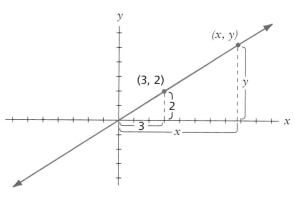

Display 3.20

This tells us that the test for a point (x, y) being on the line of Display 3.20 can be written as

$$\frac{y}{x} = \frac{2}{3}$$

Multiplying both sides of this equation by x, we get a simpler form

$$y = \frac{2}{3}x$$

The line in Display 3.20 is the set of all points (x, y) that satisfy this equation. That is, we can test to see whether or not a point is on this line by plugging its x- and y-values into the equation.

If the statement is true, the point is on the line;
if it's false, the point is not on the line.

Algebra and set-builder notation let us describe this line in a simple, easy to write way

$$\{(x, y) \mid y = \frac{2}{3}x\}$$

which is an abbreviation for

"the set of all points (x, y) such that y is $\frac{2}{3}$ of x."

1. Write the coordinates of five points that must be on the line of Display 3.20. Then write the coordinates of five points that cannot be on this line. Explain how you know that your answers are correct.

2. Pick a point somewhere on the coordinate plane and draw the line through that point and the origin. Then try to write an equation that tells you which points are on the line and which are not. When you get an equation, use it to describe the line in set-builder notation.

Do you see how you can apply the reasoning of this example to any other line through the origin? Do you recognize a pattern that you have seen before? Let us generalize what we have seen. Remember: We are talking about lines *through the origin*.

- If we know any point (c, d) besides $(0, 0)$ on such a line, (providing that $c \neq 0$).then we can find the slope of the line by dividing d by c

- The slope allows us to write as a simple equation the condition that tells us whether or not a point is on this line.
$$y = (\textbf{slope}) * x$$

- If we use the letter m to represent the slope, then the equation can be written very briefly
$$y = mx$$

- This equation determines the set of all points that make up the line. The line is
$$\{(x, y) \mid y = mx\}$$

(the set of all points (x, y) such that $y = mx$).

There is one line through the origin for which the process just described does not work. Which line is it? Explain.

The form of this equation should look familiar. It is the constant multiple pattern that you saw in the previous chapter. The constant is the ratio of the y-value to the x-value for any point on the line.

Thinking Tip

Try to generalize.
When you have solved a problem or learned an example, try to see how parts of it might apply to other similar situations.

About Symbols

In mathematics books, the letter m is the most commonly used symbol for slope (although other letters or symbols would work just as well). Let it remind you of *multiple*.

Does *every* constant multiple equation represent a line through the origin? That is, even if your friend picks a number n and doesn't tell you what it is, can you *guarantee* that the equation $y = nx$ will represent a straight line? Explain your thinking.

Problem Set: 3.3

1. A straight ramp starts at ground level and ends 2 feet above the ground after 12 horizontal feet.

   (a) What is its slope?

   (b) If you wanted to put two vertical braces, equally spaced, between the beginning and the end, where would you put them? How high would each one be?

2. Each of these pairs of points determines a line (not necessarily through the origin). Use the coordinates of the points to find the slope of each line. Then plot the points and draw the lines.

   (a) $(3, 4)$ and $(5, 9)$

   (b) $(-2, 5)$ and $(4, 7)$

   (c) $(4, 2)$ and $(1, 4)$

   (d) $(6, 6)$ and $(-1, -1)$

   (e) $(6, -6)$ and $(-6, 6)$

   (f) $(7, -4)$ and $(-3, -4)$

3. Draw coordinate axes on a piece of graph paper. For each point given here, draw the line through that point and the origin. Then find the slope of the line.

   (a) $(3, 5)$

   (b) $(5, 3)$

   (c) $(3, -5)$

   (d) $(2.8, 2.4)$

   (e) $(7.5, -4)$

   (f) $\left(\dfrac{4}{7}, \dfrac{1}{7}\right)$

   (g) $\left(\dfrac{11}{6}, \dfrac{3}{4}\right)$

4. Each of these equations describes a line through the origin. For each line, find three points that are on it and three points that are not on it.

    (a) $y = 2x$

    (b) $y = x$

    (c) $y = \frac{3}{7}x$

    (d) $y = 3.2x$

    (e) $y = -4x$

    (f) $y = -\frac{5}{2}x$

5. Match each point in Column A with the line in Column B containing it.

A	B
$(20, 6)$	$y = 0.5x$
$(0.1, 0.3)$	$y = \frac{4}{3}x$
$(50, 25)$	$y = 3x$
$(-2, 1.5)$	$y = -2x$
$(-3, -4)$	$y = \frac{-3}{4}x$
$(6, -12)$	$y = 0.3x$

6. (a) If $(3, 4)$ is a point on a vertical line, what is the x-value of the point $(x, 7)$ on that line? What is the x-value of *any* other point on that line?

    (b) In general, if a line is vertical, what can you say about its coordinates?

    (c) If a line is vertical, the definition of slope doesn't work for it. Why not?

7. Carmen is building a rectangular swimming pool in her backyard. A cross section of it, top to bottom, is shown in Display 3.21. Carmen knows that at the end of the pool where there is a low diving board, the pool must be 10 feet deep for a length of 6 feet. That end is on the left side of the diagram. She also knows that the pool can't be longer than 28 feet (because of the size of her backyard). She would like to make the shallow end only 3 feet deep. However, she wonders if that will make it too steep for someone to stand up on the slanted bottom near the shallow end.

(a) What would be the slope of the slanted bottom if the pool is 3 feet deep? Explain how you figured out your answer.

(b) What if Carmen makes the pool 4 feet deep at the shallow end? Change the diagram by redrawing it with a new bottom line. Then figure out the new slope.

(c) Which of these two designs do you think would make a better pool? Why?

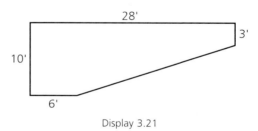

Display 3.21

8. Handy Harry is installing a rain gutter on the roof of his new house. The gutter is to be 18 feet long. The *New Complete Do-It-Yourself Manual* (by the Reader's Digest Association) tells him to slope the gutter $\frac{1}{8}$ inch for each foot of length, to allow for proper drainage. The book also tells him to install a support every 3 feet.

(a) Help Harry determine the height at which to install each support. (*Hints*: Start with a diagram. Decide where the supports go horizontally, then vertically.)

(b) Explain what this problem has to do with slope.

3.4 Exploring Lines With a Graphing Calculator

Let's review what we know about the algebra of straight lines so far:

1. Every straight line with a slope has constant slope.

2. We can find the slope of a line from any two points on it by dividing the difference in the y-values by the difference in the x-values. That is, as we go from one point on the line to another, the slope is

$$\frac{\text{change in } y}{\text{change in } x}$$

3. There is one kind of line for which this slope finding process does not work: The change in x between any two points of a *vertical* line is 0, and we can't divide by 0.

4. *If the straight line goes through the origin,* we can use $(0, 0)$ as one of the points for finding slope. Then, when we go to another point on this line, the change in y is just the y-value of that point and the change in x is just its x-value. This makes it easy to find the slope of a line *through the origin.*

5. The slope of a line is a rate of change. It tells us how many units to go up or down *for each unit we go to the right* (that is, in the positive x direction) in moving from one point to another on the line.

6. In particular, if we start at $(0, 0)$ on a line through the origin and want to go to any other point on it, then the y-value of that other point can be found by multiplying its x-value by the slope.

7. In the language of algebra, this means that the coordinates of every point on a straight line through the origin (except for the y-axis) must satisfy the equation

$$y = mx$$

where m is the slope of the line.

Statement 5 above describes how slope works in the *positive x*-direction from the origin. Explain how this statement leads to Statement 6, which applies to points in *either x*-direction from the origin.

Learning Outcomes

After studying this section, you will be able to:

Use a graphing calculator to explore the graphs of lines and to create straight line patterns

Describe how changing a number in a linear equation alters the position of its graph.

That's a pretty impressive list! We really know quite a lot already. But there are still things about the algebra of straight lines that we don't know. Here are a few:

- Do larger slope numbers correspond to steeper lines, even if the lines don't go through the origin?

- Can we always find the slope of a line from the coordinates of a single point? Can we ever do it?

- What equations describe lines that don't go through $(0, 0)$? Do these equations all look alike in any way?

How should we begin to answer these questions? One way is to think hard about the definition of slope and the ideas from Chapter 2 to see if we might find ways to generalize what we know. That's not a bad way to start, but it can be pretty frustrating if you don't see anything useful right away. Instead, let us take advantage of modern technology.

For the rest of this section we are going to go exploring with the graphing calculator.[†] Our purpose is to discover examples that might lead to interesting or useful patterns in the algebraic behavior of straight lines.

Thinking Tips

Remember these?
Find or make examples. Make a picture. Look for a pattern. Try to generalize.

EXPLORATION 1

SETUP: Your calculator has a list called **Y=** for storing equations of the form "$y = $ [**something**] ." These expressions are called functions. You will learn more about functions in Chapter 6. Find this list and clear out anything that may already be in it. Then enter the equation $y = x$ in its first line. Be sure to use the "variable" key for **X**. This first equation should look like **Y1=X**. The small 1 next to the **Y** just means that this equation is the first line that the calculator will draw.

QUESTIONS

1. Choose the Standard (or "default") **WINDOW** settings for **X** and **Y**. Now graph the line. Is the picture of $y = x$ a straight line? Should it be?

2. Do the angles between the $y = x$ line and the two axes appear to be equal? Should they be? Why or why not?

[†]These explorations were designed for the TI-82 (TI-83) calculators. The instructions may need some adjustment for other types of graphing calculators.

3. Move the cursor along the line (using the **TRACE** setting) and write down the coordinates of three points on it. How are the x- and y-coordinates of these points related to each other? Why should you not be surprised by this answer?

4. Now change the **WINDOW** to the Integer setting. Then answer questions 1, 2, and 3 again.

5. How does the Integer display of the line $y = x$ differ from the Standard display? Describe at least one way in which it is better and one way in which it is worse.

6. Which of these two **WINDOW** settings do you like better? Why?

7. Does this line have a slope? If so, what is it? If not, why not? Explain.

   Now we'll combine $y = x$ with a simple pattern to form a "family" of lines. That is, we'll make lines that have the same algebraic form and differ only by a single number. Then we'll let the calculator graph them and see if there are any interesting patterns.

EXPLORATION 2

SETUP: In the **Y=** list, enter the equations

$$y = 1x \qquad y = 2x \qquad y = 3x \qquad y = 4x$$
$$y = -1x \qquad y = -2x \qquad y = -3x \qquad y = -4x$$

Your **Y=** list should look like Display 3.22. Set the **WINDOW** to Integer coordinates and let the calculator graph the lines.

$$
\begin{array}{l}
Y_1 = 1X \\
Y_2 = 2X \\
Y_3 = 3X \\
Y_4 = 4X \\
Y_5 = -1X \\
Y_6 = -2X \\
Y_7 = -3X \\
Y_8 = -4X
\end{array}
$$

Display 3.22

QUESTIONS

1. When we changed the first equation from $y = x$ to $y = 1x$, did we change the line represented? Why or why not?

2. Copy the table in Display 3.23. Choose three integers for x and write them at the top of the third, fourth, and fifth columns. On each of the eight lines, use **TRACE** to locate the points with these x-values. Complete the table by filling in the y-values for each point.

Equation		$x =$	$x =$	$x =$	Slope
$y = 1x$	$y =$	___	___	___	___
$y = 2x$	$y =$	___	___	___	___
$y = 3x$	$y =$	___	___	___	___
$y = 4x$	$y =$	___	___	___	___
$y = -1x$	$y =$	___	___	___	___
$y = -2x$	$y =$	___	___	___	___
$y = -3x$	$y =$	___	___	___	___
$y = -4x$	$y =$	___	___	___	___

Display 3.23

3. Describe the relationship between the coordinates you found and the slopes of the lines.

4. Describe some relationship between the slope numbers and the picture. Try to state it in a way that applies to other numbers besides 1, 2, 3, 4, -1, -2, -3, and -4.

5. Change the **WINDOW** to the Standard (default) setting. How does the picture change? Do the slopes of the lines change? Explain.

6. Which **WINDOW** setting do you prefer for picturing these lines? Why?

7. Write down a question about this family of lines or a similar family of lines that this Exploration has not answered for you. Then suggest how you think you might start looking for its answer.

8. Do you see any other patterns here that you think might generalize to other lines? If so, describe them.

Design an Exploration that examines the lines

$$y = \frac{1}{1}x, \quad y = \frac{1}{2}x, \quad y = \frac{1}{3}x, \quad y = \frac{1}{4}x, \text{ etc.}$$

Use Exploration 2 to guide you in setting it up and asking appropriate questions about these lines. You may also include questions that haven't been used before, if you think they might lead to some interesting result. Then trade your Exploration with someone else in the class and try to work through the one you get.

So far, we have been changing the equation $y = x$ just by multiplying x by various numbers. What if we add or subtract, instead?

EXPLORATION 3

SETUP: In the **Y=** list, enter the eight equations formed by adding or subtracting 1, 2, 3, 4, as shown in Display 3.24. Look at the graphs of these equations, using both the Integer and the Standard **WINDOW** settings.

$$
\begin{array}{l}
Y_1 = X + 1 \\
Y_2 = X + 2 \\
Y_3 = X + 3 \\
Y_4 = X + 4 \\
Y_5 = X - 1 \\
Y_6 = X - 2 \\
Y_7 = X - 3 \\
Y_8 = X - 4 \\
\end{array}
$$

Display 3.24

QUESTIONS

1. Describe in words what happens to the line when you add a positive number to the right side of the equation $y = x$. Use your description to predict where the line $y = x + 25$ would be. Check your prediction with your calculator, if you can.

2. Describe in words what happens to the line when you subtract a positive number from the right side of the equation $y = x$. Use your description to predict where the line $y = x - 25$ would be. Check your prediction with your calculator, if you can.

3. Is subtracting a positive number the same as adding a negative number? Try to restate your answers to questions 1 and 2 in the form of a single principle.

4. Are the slopes of these eight lines all the same, or are they different? Defend your answer.

5. How could you change these equations so that all the lines slope downward from left to right? Try it.

6. State at least one other question that this **Exploration** suggests to you.

Now we'll try putting all the pieces together. We'll start with a slightly more complicated family of lines. See if you can predict what the picture will look like *before* you let the calculator draw the graphs.

EXPLORATION 4

SETUP: Choose the Integer setting for the coordinate system. Enter these eight expressions in the first eight lines of **Y=**

$$x, .75x, .5x, .25x, -.25x, -.5x, -.75x, -x$$

QUESTIONS

1. *Before* you graph these lines, make a rough sketch of how you think the picture will look. If you can't even begin to do this, write down a specific question about what you think is getting in your way.

2. Graph the eight lines with your calculator. Were you right? Close?

3. What do you think will happen to the graphs if you add 5 to the right side of every equation in the **Y=** list?

4. Do it; add 5 to the right side of every equation in the **Y=** list. Then make a copy of the table in Display 3.25. Use **TRACE** to find on each line the point with *x*-value 12, and put this information into the table. In the far left column, write the slope of each line. How (if at all) are the coordinates of these points related to the slopes of their lines?

Equation	x	y	Slope
$y = x + 5$	12	_____	_____
$y = .75x + 5$	12	_____	_____
$y = .5x + 5$	12	_____	_____
$y = .25x + 5$	12	_____	_____
$y = -.25x + 5$	12	_____	_____
$y = -.5x + 5$	12	_____	_____
$y = -.75x + 5$	12	_____	_____
$y = -x + 5$	12	_____	_____

Display 3.25

5. If you think about the pattern of these lines in a certain way, a "middle line" seems to be missing. Can you describe the missing line? What is its slope? What is an equation for it? See if you have the right equation by adding it to your Y= list and graphing all the lines together.

6. Can you generalize your observations in any way? Do these results raise any questions for you? If so, what are they?

Let's look at one more family of equations. These are suggested by the fact that exponentiation is an operation of algebra. What happens if we go back to the equation $y = x$ and use 1, 2, 3, 4, etc. as exponents for x?

EXPLORATION 5
SETUP: Enter the equations

$$y = x^1 \qquad y = x^2 \qquad y = x^3 \qquad y = x^4$$

into the first four lines of the Y= list. Clear the other lines. Look at the graphs of these equations using the Standard and Integer **WINDOW** settings. Also look at them with the setting that the TI calculators call Decimal. If you don't have a setting of this name on your calculator, go to the Integer setting and divide each of the **max** and **min** values by 10.

Now go back to the **Y=** list and turn off these four equations. Into the next four lines of the list, enter the equations

$$y = x^0 \qquad y = x^{-1} \qquad y = x^{-2} \qquad y = x^{-3}$$

Look at the graphs of these equations using the Standard, Integer, and Decimal **WINDOW** settings.

QUESTIONS

1. What makes these sets of equations different from all the others we have been examining in this section?

2. Do all of these graphs have slope? Do any of them have slope? Why or why not?

3. Which **WINDOW** setting do you like best for these families of lines? Why?

4. Think of at least one question that these displays suggest to you, and write it down as specifically as you can. Then suggest a first step in finding an answer to it.

5. What other exponents do you think might give us interesting families of graphs?

As **Exploration 5** suggests, equations in which variables are raised to powers other than 1 or 0 describe sets of points that are not simply straight lines. We leave the study of such equations and their pictures to another time. For now, we shall study only sets of points (x, y) with equations that can be put in the form

$$y = [\text{some number}] * x + [\text{some number}]$$

As you learned in Chapter 2, algebra allows us to replace clumsy expressions like "[**some number**]" with single letters that mean the same thing. That is, in the language of algebra, this equation form is written as

$$y = a * x + b$$

In this expression, a and b stand for fixed numbers (constants). Different letters are used to show that the two numbers might be (but don't have to be) different.

Now, it turns out that the set of points described by *any* equation of this form is a straight line. You have already seen evidence of this in the Explorations we have done; the next section will show you more about it. This is a very important fact, so let us take the time to state it again, a little more carefully.

> **A Fact to Know:** If an equation has the form $y = ax + b$, where a and b are constants, then the points (x, y) that satisfy it form a straight line.

For this reason, any equation that can be put in the form $y = ax + b$ is called a **linear equation**.

1. **List five examples of linear equations. For each one, say what a and b are.**

2. Is $y = \frac{2}{3}$ a linear equation? If so, what are a and b? If not, why not?

3. Is $y = 5$ a linear equation? If so, what are a and b? If not, why not?

4. Is $3x + 2y = 4$ a linear equation? If so, what are a and b? If not, why not?

Before ending this section, let us clear up one custom of word usage. Mathematicians usually use the word *line* to mean *straight line*. Sometimes the whole phrase *straight line* is used just to emphasize the straightness. The word they use for a line that may or may not be straight is *curve*. We shall adopt this custom from now on because it is simple and efficient.

When you see the word *line*, assume that it means *straight line*, unless you are told otherwise.

Problem Set: 3.4

Each of problems 1 through 8 has two parts

(a) Using the Integer screen setting, enter equations in the Y= list that form the pattern shown here. The scale marks have been left out deliberately to allow you to choose what works best for you.

(b) After you have the display the way you want it, look at it with the Decimal setting and then with the Standard setting. Make a note of any interesting changes you see.

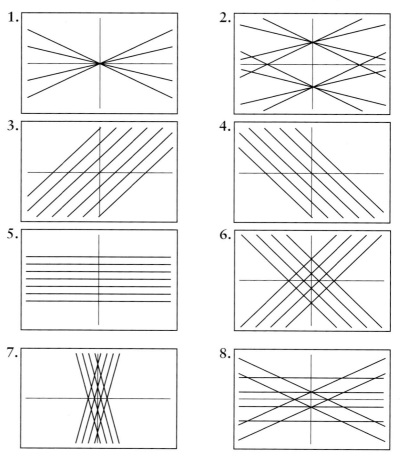

9. Design and produce a straight line design of your own on the calculator. See how creative you can be!

10. Why do you suppose you weren't asked to make this picture?

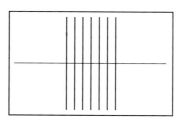

3.5 Lines That Do Not Go Through (0, 0)

The ramp at the Fuzzy Friends Toy Co. outlet store was a great success with their customers except when rain storms came. Then the rain made it too slippery to use. The company decided to put a roof on it, 7 feet above the ramp and parallel to it (Display 3.26). The poles needed to hold up the roof were to be placed 4 feet apart, next to each ramp support, fastened at the bottom to the surface of the parking lot. How long should they have cut each of these poles?

Learning Outcomes

After studying this section, you will be able to:

Use the slope and the y-intercept of a line to write an equation for it

Draw lines described by equations of the form $y = mx + b$

Use the equations of two lines to determine whether the lines are parallel.

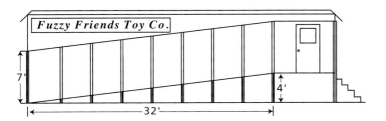

Display 3.26

Can you begin to answer this question by drawing a simpler picture? Here are the important facts that the picture should show:

- The roof is to be 7 feet above the ramp all the way up.
- Each pole is to be set right next to a ramp support.
- The bottom of each pole is to rest on the parking lot surface.

One way to draw such a picture is by making two slanted lines—one for the ramp and one for the roof. We have done this for you in Display 3.27. Compare this with Display 3.16. The y-values of the points on the lower slanted line show the height of each ramp support.

1. The missing y-values of the points on the upper slanted line should show the height of each roof pole. Copy these nine pairs and fill in their missing y-values.

2. How are the heights of the poles related to the heights of the ramp supports?

3. The ramp has a slope of $\frac{1}{8}$ because it rises one foot for every eight horizontal feet. If the roof is 7 feet higher than the ramp all the way up, what is the slope of the roof?

4. What equation represents the points on the ramp line?

5. What equation represents the points on the roof line?

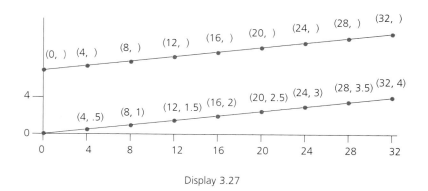

Display 3.27

This ramp-and-roof story leads us to an important fact. Two lines that are parallel (the same distance apart everywhere) must have the same slant. That is, they must rise the same amount for each unit of their length. In other words:

Parallel lines have the same slope.

This simple observation is the key to understanding equations of straight lines that do not go through the origin.

Let's look at an example. Here is an equation for a line that does not go through (0, 0):

$$y = 2x + 3$$

Be suspicious! Should you believe us when we say that this equation describes a line that

(a) is straight and
(b) doesn't go through (0, 0)?

Check by putting this equation into your calculator and graphing it.

If you understand that the "sawtooth" picture you get is just the calculator's way of showing straightness with a limited number of screen dots, then you can see that this is a straight line, and it doesn't even come close to going through (0, 0).

Of course, you didn't have to make a picture to know that the graph of this equation doesn't contain (0, 0). If (0, 0) were a point of this graph, its x- and y-values would have to make the equation a true statement. Now, when 0 and 0 are put in for x and y, we have

$$0 = 2 * 0 + 3$$

But $2 * 0 = 0$ and $0 + 3 = 3$, so this says $0 = 3$, which is nonsense. So (0, 0) can't be on this line.

What is the slope of this line? Would you believe us if we told you that it's 2? Why should you believe this? An answer that's *not* very good is: "Because the book says so, and textbook authors never make a mistake!" Here's one way to find out for yourself:

1. Start by remembering what *slope* means. The slope of a line can be found by taking any two points on it and dividing the difference in the y-values by the difference in the x-values.

2. Pick a number (any number you want) for an x-value. Call it x_1. The subscript 1 next to x just means that this is the first x-value you picked.

3. Find the y-value that goes with the x-value you chose. Put x_1 into the equation $y = 2x + 3$ and calculate y. Call it y_1.

4. Pick another x-value and call it x_2. Then find y_2, the y-value that goes with your choice for x_2.

5. Find the difference in the y-values and write it down like this

$$y_1 - y_2 = \underline{}$$

6. Find the difference in the x-values and write it down like this

$$x_1 - x_2 = \underline{}$$

Thinking Tip

Check the definitions. If you can't recall exactly what a word means, look it up.

7. Divide the difference in the *y*-values by the difference in the *x*-values. We don't know which numbers you picked, but we are so confident in your work (and in the idea of slope) that we have written your answer for you.

$$\frac{y_1 - y_2}{x_1 - x_2} = 2$$

Are we correct?

8. What if you calculate the *y* difference as $y_2 - y_1$ and the *x* difference as $x_2 - x_1$? Do you get the same value for the slope? Try it.

9. What if you calculate the *y* difference as $y_2 - y_1$ and the *x* difference as $x_1 - x_2$? Do you get the same value for the slope? Try it.

Now that you know the slope of the line $y = 2x + 3$, it is easy to write an equation for the line through (0, 0) that has the same slope. We have already seen that equations for lines through (0, 0)—except the *y*-axis—all have the form

$$y = (\text{slope}) * x$$

so the equation

$$y = 2x$$

must describe the line through the origin that is parallel to $y = 2x + 3$. (Put $y = 2x$ into your calculator and have it graph both equations at once.)

What's the difference between these equations? The short answer is "3." But what does that really mean? Think of it this way.

The line described by $y = 2x$ is the set of all points whose *y*-values are twice their *x*-values. To get from each of these points to the line $y = 2x + 3$, we just have to add 3 to each of these *y*-values.

In the same way, if we want to move the line $y = 2x$ up by 100, we just add 100 to each *y*-value. That is, we make $y = 2x + 100$.

1. What equation describes the line $y = 2x$ moved up by 25? By 3.724? By $\frac{4}{7}$?

2. What equation describes the line $y = 2x$ moved down by 25? By 3.724? By $\frac{4}{7}$?

3. Generalize your answers to parts 1 and 2 by making up a way to describe the equation of *any* line that is parallel to $y = 2x$.

4. Graph your equations from parts 1 and 2 on your calculator. Find the points where each line crosses the *y*-axis. How are the coordinates of these points related to the equations of the lines?

You can think of *all* lines that are parallel to $y = 2x$, up and down the entire plane, as represented by equations of the form

$$y = 2x + \boxed{}$$

where the box can be any number. This is usually written as

$$y = 2x + b$$

You pick out one of these many lines by choosing a number for *b* (the box). The number you choose tells you the point at which your line crosses the *y*-axis. For this reason it is called the **y-intercept**.

Of course, what we have said here about lines with slope 2 could have been said about lines with any slope. Any number, positive or negative, can be used as the slope of some line through the origin. You can think of *all* lines through (0, 0) — except the *y*-axis — as represented by equations of the form

$$y = \boxed{} * x$$

where the box can be any number. This is usually written as

$$y = mx \quad \text{or} \quad y = ax$$

but other letters will work just as well. Whatever letter you use stands for the slope of the line. It is an example of a *coefficient*. When a variable is multiplied by a number in an algebraic expression, the number is called a **coefficient** of the variable.

The lines of the form $y = mx$ for all possible values of *m* form an infinite family of lines that go through (0, 0). Think of (0, 0) as the nose of a cat with infinitely many whiskers! Choosing a particular number for *m* determines whether the

slant of the line is steep or gentle. The sign of the slope number determines whether the line slants up or down as it goes from left to right.

Get your calculator to graph these lines on the same axes.

$$y = 2x \quad y = .5x \quad y = -3x \quad y = -.4x \quad y = 20x$$

along with any other three lines through the origin that you choose. Which one is the steepest? Which ones slant downward from left to right?

(*Hint*: TRACE can help you keep track of which equation goes with which line.)

It's time to summarize what you have seen about the algebra of straight lines.

- Saying that an equation describes a line means that the line is the set of all points whose coordinates make the equation true.

- Except for vertical lines, every straight line in the plane can be described by an equation of the form $y = mx + b$, where x and y are variables that stand for the coordinates of the points, and m and b are constants (fixed numbers).

- Every equation of the form $y = mx + b$ describes a straight line. The number m is its slope and the number b is its y-intercept. Equations that can be put in this form are called *linear equations*. This form of a linear equation is called its **slope-intercept form**.

- You can write an equation to describe a particular line by finding two numbers: the slope and the y-intercept of the line.

  - The slope tells you the steepness of the line and whether it is rising or falling as it goes from left to right.

  - The y-intercept tells you where the line crosses the vertical axis.

a

Which of the Explorations in Section 3.4 gave you the clearest idea of how the *y*-intercept of an equation relates to the picture of the line? Which Exploration gave you the clearest idea of how the slope relates to the picture of the line? Explain.

What do you suppose is meant by the "*x*-intercept"? How would you find it from the graph of a line? How would you find it from the equation of a line? Find the *x*-intercept of the line $y = 2x + 3$.

b

Some people say that vertical lines have infinite slope; some say that they have no slope. Both descriptions are accurate, in a way, but neither one is perfect. Write a paragraph describing what is good and what is bad about each of these ways of describing vertical lines.

Problem Set: 3.5

1. The equation $y = \frac{1}{2}x - 3$ represents a line.

    (a) Write a sentence or two to describe the graph of this line.

    (b) Is $(4, -1)$ on this line? How do you know?

    (c) Is $(0, 6)$ on this line? How do you know?

    (d) Find the coordinates of three other points on this line.

    (e) Graph the line on your calculator.

    (f) Use **TRACE** to check your answers to parts (b), (c), and (d).

2. Write an equation for a line that is parallel to $y = -3x + 5$ and below it. Check your answer by graphing both lines on your calculator.

3. (a) Write an equation for a line that has *y*-intercept $(0, 0)$ and contains the point $(-5, 3)$.

    (b) Is $(6, -3.6)$ on this line? Why or why not?

    (c) Is $(4, 2.4)$ on this line? Why or why not?

    (d) Write an equation for a line that has *y*-intercept $(0, -2)$ and is parallel to the line in part (a).

4. (a) Write an equation for the line with slope 0 and y-intercept 3. Then graph the line.

   (b) Write an equation for the line that is parallel to the line in part (a) and goes through (-2, 5). Check your answer by graphing both lines on your calculator.

5. Cheryl babysits for one family regularly. She is paid $2.75 for every hour she babysits. Let's look at a graph of the relationship between how much she earns per job, E, and how many hours she works, h. Start by drawing coordinate axes on a piece of graph paper. Label the vertical axis E and the horizontal axis h.

   (a) If Cheryl works for one hour, she earns $2.75. If she works for two hours, she earns $5.50. If she works for three hours, how much does she earn? Write down and then plot the three ordered pairs that represent this information.

   (b) Each time Cheryl works another hour, what happens to her earnings? What is the pattern here?

   (c) Draw a line through the three points you have plotted. What is its slope? In terms of Cheryl's job, what does the slope represent?

   (d) Can you estimate how much Cheryl will earn if she works 7.5 hours? If you can, do it. If you cannot, explain why you think it can't be done.

   (e) Write an equation that represents Cheryl's earnings in terms of the number of hours she works.

   (f) Does it make sense to discuss how much Cheryl earns when she works -2 hours? What does the answer to this question suggest about the graph of the line of part (e)?

   (g) If Cheryl has someone drive her to and from the babysitting job, instead of the family she is sitting for, she is paid $4 in addition to her hourly rate. How does this change the equation that represents her earnings? How does it change the graph? Graph this new line on your calculator.

6. Not everyone uses the same symbols for the slope and the *y*-intercept when describing linear equations. For instance, one of the statistics menus of the TI-82 (TI-83) calculators uses these two forms:

$$y = ax + b \quad \text{and} \quad y = a + bx$$

where *a* and *b* are constants.

(a) In $y = ax + b$, what stands for the slope and what stands for the *y*-intercept? How do you know?

(b) In $y = a + bx$, what stands for the slope and what stands for the *y*-intercept? How do you know?

7. This problem refers back to an example of Chapter 2. In that example, an electric company computed the total monthly charge (*T*) for each customer by multiplying the price per kiloWatt-hour ($0.09) by the number of kiloWatt-hours used (*u*) and adding a service charge of $8.50. The resulting equation is

$$T = 0.09u + 8.50$$

If we think of the horizontal axis as the *u*-axis and the vertical axis as the *T*-axis, this equation describes a straight line.

(a) What is its slope?

(b) What is its *T*-intercept? That is, where does it cross the *T*-axis?

(c) Draw a graph of this line on a piece of paper.

(d) Display this line on your graphing calculator.

(e) Find two positive numbers that are the coordinates of a point on this line. Then explain, in terms of the electric bill, what it means to say that the point is on this line.

(f) Is the point (−1, 8.41) on this line? Can you interpret this in terms of an electric bill? If so, how? If not, why not?

8. Chapter 2 also described another rate schedule for the electric company, calculated by the linear equation

$$T = 0.10u + 6.00$$

(a) What is the slope of this line? What part of the electric bill does it describe?

(b) What is the *T*-intercept of this line? What part of the electric bill does it describe?

(c) Compare this line with the line of problem 7. Start by graphing them on the same set of axes (either on paper or with your calculator). Which line is steeper?

(d) Are the lines getting closer together or farther apart as they go to the right? In terms of the cost of electricity, what does this mean?

(e) Do you think that these lines will ever cross? Explain.

9. It is late June. Bobby has saved $2,500 for his summer vacation, which started on Sunday, May 29, and ends on Saturday, August 27. His average spending for the first four weeks has been $240 a week.

(a) Write an equation showing the amount A of money that Bobby will have left after W weeks of his summer vacation, assuming that he continues to spend at the same rate.

(b) What are the slope and the y-intercept of the line for this equation?

(c) Graph the line.

(d) Will Bobby have enough money to make it through the summer? If not, in which week will he run out of money?

(e) What is the most money that Bobby could spend per week to make it through the summer? Write an equation that represents this new spending pattern, and graph it on the same axes with the original equation. State your answer to the nearest dollar.

10. Set your graphing calculator Window so that $1 \leq x \leq 5$ and $1000 \leq y \leq 1500$. Then graph these two equations:

$$y = 1000 + 70x \qquad\qquad y = 1000 * 1.07^x$$

These two equations come from Chapter 2. They represent simple vs. compound interest on Alfredo's $1,000 gift. Is either of these lines straight? Are they both straight? Use the definition of slope to help you justify your answers. (*Hint*: Remember that the slope of a straight line is the same between *any* two points on the line.)

3.6 Two Points Determine a Line

The title of this section is a statement that you may have heard before. It's a commonsense geometric idea taken from everyday experience: You can draw exactly one straight path between two dots on a piece of paper.

But is it still true if we're talking about lines in *algebra*? In this chapter we have seen that lines can be described by equations. If we know only the coordinates of two points, can we always find an equation for the line that goes through them?

What do *you* think? How would you start to handle this question? What Thinking Tip might be useful here?

An equation for a line tells us how the *x*- and *y*-coordinates of the points on that line are related. For instance, saying that a line has equation $y = 2x + 1$ means that, for any point on this line, if you double its *x*-coordinate and add 1, you will get its *y*-coordinate.

In the preceding section, you saw how to make an equation for a line from its slope and its *y*-intercept. For instance, if we know that a line has slope 5 and *y*-intercept 3, we can write down an equation for that line: $y = 5x + 3$. In general, such an equation for a line has the form

$$y = [\text{slope}]x + [y\text{-intercept}]$$

So the question of finding an equation for a line comes apart into two pieces.

A. Can we find the slope?

B. Can we find the *y*-intercept?

Let's try an example. How about finding an equation for the line through (1, 8) and (3, 14)?

A. Can we find the slope? Of course! The slope is the change in *y*-values divided by the change in *x*-values *between any two points on the line*, so we can use these two points to find the slope. We can use the points in either order, so let's

Learning Outcomes

After studying this section, you will be able to:

Use the coordinates of two points to write an equation for the line through those points

Use graphs and equations of lines to analyze rate of change in various real world situations.

take (3, 14) first. Then the slope is

$$\frac{14 - 8}{3 - 1} = \frac{6}{2} = 3$$

That was easy! Now we know that the equation can be written in the form

$$y = 3x + b$$

for some number b. How do we find b?

B. Let's see what we know about b. First of all, we know that b is a constant. That is, it's a single number that doesn't change, no matter where we are on the line. We also know that every point on the line satisfies the equation $y = 3x + b$. Since (1, 8) is a point on the line, we know that

$$8 = 3 * 1 + b$$

must be true. That is,

$$8 = 3 + b$$

Now we find b by subtracting 3 from both sides of the equation. In other words, 8 is 3 more than b, so b must be 5.

Putting the results of **A.** and **B.** together, the equation of the line is

$$y = 3x + 5$$

1. If we take (1, 8) as the first point, do we get the same slope? Explain.

2. If we put (3, 14) into the equation $y = 3x + b$, do we get the same y-intercept? Try it.

3. What equation describes the line through (1, 8) and (3, 14)?

4. Is the point (5, 20) on this line? Why or why not?

5. Is (8, 14) on this line? Why or why not?

6. Is (−3, −4) on this line? Why or why not?

Can we generalize the method of this example so that we always get an equation from two points? That is, can we create an *algorithm* for finding an equation from two points? If we write out the steps without using any of the specific numbers from the example, then the method should work every time, right? Here are the steps.

1. Use the definition of slope (change in y divided by change in x) to find the slope from the coordinates of the two points.

2. In the equation form $y = mx + b$, replace m by the slope value you just found, and replace x and y by the coordinates of one of the points.

3. Find b by solving your equation in 2. for b.

    1. **Find an equation for the line through the points (2, 5) and (6, 3).**

    2. **Recall that the slope of a line tells you how to get from one point on the line to another, as follows. The (positive) denominator tells you how many units to move to the right; the numerator tells you how many units to move up or down. Use this process to find three more points on this line. Where should you start?**

    3. **What if you reverse the direction of part 2? That is, what if you move to the left? How must you adjust the way you move up or down? Use this process to find three more points on this line.**

    4. **Draw a graph of this line on a piece of paper. On your graph, mark the two points of part 1 and all the points you found in parts 2 and 3. Label all these points with their coordinate pairs.**

Linear equations are used to describe many situations in the real world. Sometimes the situations don't appear at first to have anything to do with straight lines or points on a plane. But slope just represents a constant rate of change between two variables (things that can take on different numerical values). In the rest of this section, we look at an example of using slope and the equation for a line determined by two points to describe

Ways of Measuring Temperature

Most of the world uses the metric system to measure things—except in the U.S., where we use the English system. Do you know what they use in England? No, they don't use the English system; they use the metric system! One thing we measure is temperature. If you can't keep track of temperature, then you can't build or drive a car, keep yourself healthy, launch a rocket, build a refrigerator, or cook anything fancier than burnt toast.

The English temperature scale is called **Fahrenheit**. On this scale, a comfortable house is at 70°, dishwashers use 140° water, and the mean temperature of a healthy human body is about 98°.

Estimate the Fahrenheit temperature of:

1. a cup of hot soup

2. a glass of cold milk

3. the coldest day of the year in your town

4. the hottest day of the year in your town

5. the freezing point of water

6. the boiling point of water.

Can you think of some other temperatures that are useful to know?

Two of the most important temperatures for science and technology are the places on the scale where water freezes and where it boils. Water probably is the most important compound on Earth. No other naturally occurring compound does more for us. Water is even responsible for most of the soil on the planet. Just before freezing solid, water expands. In this way, freezing water can break boulders into rocks, rocks into pebbles, and pebbles into sand. It acts as a universal solvent for all the reactions that take place in the atmosphere and in the ocean, in our bodies and in every other living thing. Thus, the places on a temperature scale that measure the freezing and the boiling of water are useful numbers to know.

You probably have heard of a metric temperature scale known as **Celsius**. On this scale, your comfortable living room measures 21°, the dishwasher water is 60°, and the human body, when healthy, averages 37°. Water temperature can also be measured on this scale. When it freezes, the Celsius scale reads 0°, and when it boils, the Celsius scale reads 100°. These are much easier numbers to deal with, so they make convenient anchor points for the scale.

The Celsius scale is also known as *centigrade*. Why is this word used? (*Hint*: Try breaking the word in half and thinking about the meaning of each part separately.)

Has somebody already told you that there is a formula for converting from Fahrenheit to Celsius? Do you remember it? Do you know where it comes from? Let's see.

You know how to use the coordinates of two points to get an equation for the straight line that goes through them. Now, the two important points on both temperature scales are the freezing point and the boiling point for water. When water freezes, we can measure it with either scale. The scale doesn't affect the water; it's still the same freezing water, no matter which scale we measure it with. So we know that 32° Fahrenheit and 0° Celsius label the same point in the possible temperatures of water.

If we had written "32°F and 0°C" in the preceding sentence, would you have known what we meant? Do you think these abbreviations are useful here? What is possibly confusing in interpreting C? Does it matter? Why or why not?

We also know that 212°F and 100°C represent the same temperature, the boiling point of water. Let's set up coordinate axes to represent Fahrenheit and Celsius, and plot the points (32, 0) and (212, 100). Then we'll draw the line connecting them (Display 3.28.)

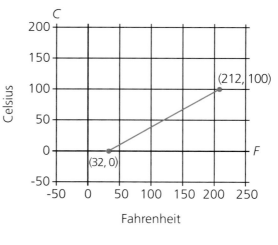

Display 3.28

To get the slope of this line, we need to remember two things:

- The slope equals the change in y between these two points divided by the change in x between these two points.

- The labels x-axis and y-axis are just abbreviations for horizontal axis and vertical axis.

1. Find the slope of the line segment in Display 3.28. Explain your method.

2. If you extend this line segment, it will cross the vertical axis (the C-axis). Set up an equation relating F (the number of degrees Fahrenheit) to C (the number of degrees Celsius) that will enable you to find the C-intercept.

Did you get $\frac{5}{9}$ for the slope? Good. This means that all points on the line fit the equation

$$C = \frac{5}{9}F + b$$

You also know two points on this line, (32, 0) and (212, 100).

1. Describe how to use one of these points to find b; then do it.

2. Which point did you use? Find b using the other point. Do you get the same value?

3. In this case, is one point easier to use than the other? If so, why?

4. To the nearest degree, what is the Celsius temperature when a Fahrenheit thermometer says 0°?

At this stage, you should have the equation

$$C = \frac{5}{9}F - \frac{160}{9}$$

This is an algorithm (a formula) for converting *any* temperature from Fahrenheit scale to Celsius scale. You also have a graph (Display 3.28) that can be used to estimate conversions quickly. Try out both of these tools on the following questions.

1. Use the graph to estimate the Celsius temperature when the temperature is 100°F. Then use the equation to find the answer to the nearest degree.

2. Use the graph to estimate the Celsius temperature when the temperature is 50°F. Then use the equation to find the answer to the nearest degree.

3. Use the graph to estimate the Fahrenheit temperature when the temperature is 50°C. Then use the equation to find the answer to the nearest degree.

4. Which method is easier, the equation or the graph? Which is quicker? Which is more accurate?

1. How would you convert from Celsius to Fahrenheit? Find an equation for doing this. Draw the graph of this equation. Then explain your work.

2. Describe any connection you see between the C-to-F equation you found here and the F-to-C equation we found earlier.

3. Describe any connection you see between the C-to-F graph you drew here and the F-to-C graph of Display 3.28.

We usually convert from one temperature scale to another only because we might need to communicate findings or questions to people who use different scales. People need to know the temperature of a thing, whether it's the air outside, the oil in their car engine, or the milk in the refrigerator.

Temperature is just a way to tell if an object is going to lose or gain heat when it is near any other objects. Heat always flows from a higher temperature place to a lower temperature place. Heat flows from the higher temperature ear of corn, for instance, to your lower temperature finger, and it also flows from your higher temperature finger to the lower temperature ice cube. Keep in mind that an object's temperature doesn't change just because you measure it with a different scale, any more than you grow taller when you are measured in centimeters instead of inches.

Scientists have another scale for temperature; it is called **Kelvin**.[†] The Kelvin scale has degrees that are the same size as Celsius degrees (there are one hundred of them between freezing and boiling water) but it has a different "anchor." Water is important here on earth, but once we started exploring space and other planets and stars, it made sense to think about all the chemicals we might find; our scale based on water might be too limited. Scientists decided that zero really ought to be the place on the scale where an object has lost all the heat it can—the molecules themselves have stopped moving. This is called **absolute zero**, and it is 273.2 degrees below the temperature of freezing water. It is very, very difficult to keep something's temperature near absolute zero. Even the space between the stars is not completely cold.

1. If water freezes at 273.2 degrees on the Kelvin scale and boils at 373.2 degrees, what is the conversion between Celsius and Kelvin? Use the information in the paragraph just before these questions to write an equation for conversion from Celsius to Kelvin. Explain how you got it.

2. Does your expression describe a straight line? If so, what are its slope and its *K*-intercept?

REFLECT

This chapter has been about straight lines and algebra. You learned how to describe straight lines in a plane with algebraic equations by using a coordinate system. Using two coordinate axes, a horizontal one and a vertical one, you can label each point of the plane with an ordered pair of numbers that tells

---

[†] It is named after the British mathematician and physicist William Thomson (1824–1907), 1st Baron Kelvin of Largs.

you where the point is. Each nonvertical straight line in a coordinate plane determines two specific numbers: its slope and its *y*-intercept. Its slope measures how steeply the line rises or falls with respect to the horizontal axis. Its *y*-intercept tells you where the line crosses the vertical axis.

The slope, *m*, and the *y*-intercept, *b*, of a straight line can be used to write a simple algebraic description of all the points on that line. A point (x, y) is on that line if and only if its coordinates make this equation true.

$$y = mx + b$$

Every nonvertical straight line can be described by an equation of this form. Moreover, each equation of the form

$$y = ax + b$$

describes a straight line with slope *a* and *y*-intercept *b*. For this reason, such equations are called *linear equations*. (Vertical lines are described by equations of the form $x = c$, where *c* is the point at which the line crosses the *x*-axis. Slope of vertical lines is undefined because the change in *x* is zero.)

Problem Set: 3.6

1. For each part, find an equation that describes the line containing the two given points.

   (a) $(2, 11)$ and $(5, 17)$ (e) $(4, 3)$ and $(4, 7)$

   (b) $(2, 3)$ and $(5, 8)$ (f) $(-6, 2)$ and $(1, -8)$

   (c) $(6, 2)$ and $(1, 8)$ (g) $(\frac{1}{2}, 3)$ and $(-\frac{1}{2}, 2)$

   (d) $(3, 4)$ and $(7, 4)$ (h) $(\frac{1}{4}, -\frac{1}{2})$ and $(3\frac{1}{4}, -2\frac{1}{2})$

2. Imagine yourself standing on a huge coordinate system. You move carefully from one point to another.

   (a) If you move so that the *x*-coordinate of the point you are standing on is always 3 more than the *y*-coordinate of that point, will you be moving in a straight line? (*Hint*: Draw a picture.)

   (b) Write down the coordinates of two possible points on which you could stand if you satisfy the condition of part (a).

(c) Write the equation of the line through those two points.

(d) Is more than one line possible? Explain.

3. Marguerita and Edwin's math teacher decided to scale the last set of test scores by adding the same number of points to each student's grade. All questions on the test were worth the same number of points. Edwin had 18 correct answers, which gave him a scaled grade of 70. Marguerita had 24 correct answers, giving her a scaled grade of 90.

   (a) Write an equation that computes the new test scores from the number of correct answers.

   (b) Describe what the slope and y-intercept represent in the equation.

   (c) Draw a graph of the line.

   (d) If Teresa had 20 correct answers, what would her scaled grade be?

4. Casey lives 15 miles from me. My brother gives me a ride to the park, which is 2 miles from my house and 2 miles closer to Casey's house. Then I ride my bike at an average of 5.5 miles per hour to Casey's house.

   (a) Write an equation for the distance I am from my house in terms of the number of hours, h, that I have been riding.

   (b) How far will I be from my house in one hour?

   (c) In terms of the equation, what does the 5.5 represent?

   (d) In terms of the graph, what does the 5.5 represent?

   (e) Does it make sense to let $h = -5$? Explain.

   (f) Does it make sense to let $h = 1.75$? Explain.

   (g) Does it make sense to let $h = 5$? Explain.

5. Dr. Dull gave a 90-minute lecture in the assembly hall last
   week. When his lecture began, there were 102 people in the
   hall. Ten minutes later there were 87 people in the hall.
   After 20 minutes there were 72 people left in the hall. After
   30 minutes there were 57 people left in the hall. After
   40 minutes there were 42 people left in the hall.

   (a) Write an equation for the number of people left in the
       hall in terms of the number of minutes since Dr. Dull
       began his lecture. Use P for the number of people and
       t for the time in minutes.

   (b) What does the y-intercept represent in this situation?

   (c) Does it make sense to talk about the number of people
       in the lecture hall 20 minutes before his lecture began?
       If so, do you think that this number will fit the equation
       you wrote? Why or why not?

   (d) Can you use your equation to predict the number of
       people in the room after one hour? What must you
       assume?

   (e) Can you use your equation to predict the exact number
       of people in the room after 45 minutes? Explain why or
       why not.

   (f) When will the room be empty? What point on the
       graph helps you answer this question?

6. When you attach a weight to the end of a spring, it
   stretches. The length of a spring with a weight suspended
   from it depends on the length of the spring without the
   weight, the size of the weight, and the elasticity of the

spring. The elasticity of a spring depends on the material of which it is made and on its size, shape, and thickness. The length of the spring increases at a constant rate as the weight increases. Many scales use this principle in measuring weights.

Allison and Lonnie experimented with a spring in physics class. They collected the following data:

Weight hung on spring, in grams:	10	15	20	25	30
Length of the spring, in cm:	22	23	24	25	26

(a) Write an equation for the length of the spring in terms of the weight suspended from it.

(b) What is the length of the spring with nothing attached? Explain how you arrived at your answer.

(c) The slope of the line is sometimes called the *constant of elasticity* of the spring. What is that constant for Allison and Lonnie's spring?

(d) Do you think your equation would work if they suspended a 100-gram weight from their spring? Explain.

If you want to try a similar experiment, a rubber band works on the same principle. What happens to a rubber band if you hang a very heavy weight on it?

7. This problem is about **density**, which is the weight— actually, it's mass, but as long as we remain on one planet, the numbers for weight and mass are the same—divided by the volume (the space that the object takes up). That is,

$$\text{density} = \frac{\text{mass}}{\text{volume}}$$

Water has a density of 1 gram per milliliter (g/mL). That is, 1 milliliter (mL) of water weighs 1 gram (g), 2 mL weigh 2 g, 100 mL weigh 100 g, etc. Things that have a density greater than 1 g/mL will sink in water; anything with a density less than 1 g/mL will float.

Display 3.29 lists a dozen things you found in the attic, along with their masses (in grams) and volumes (in milliliters). Display 3.30 shows each of these things as a point on a graph with mass (in grams) on its horizontal axis and volume (in milliliters) on its vertical axis. Your teacher will give you a copy of Display 3.30.

Object	Mass	Volume	Density
A	1	10	0.1
B	1	4	0.25
C	4	2	2
D	1	3	0.333
E	3	10	0.3
F	10	3	3.333
G	7	3	2.333
H	9	10	0.9
I	3	6	0.5
J	7	6	1.1667
K	2	3	0.6667
L	6	6	1

Display 3.29

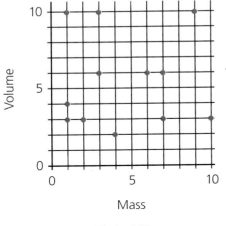

Display 3.30

(a) On your copy of Display 3.30, label each point with the correct letter from the table in Display 3.29.

(b) Since water has a density of 1 g/mL, 2 g of water take up 2 mL, 10 g take up 10 mL, etc. Use this information to graph on your copy of Display 3.30 the line that describes water.

(c) Which of your dozen objects is made of water?

(d) Which of these objects will float in water? Which will sink?

(e) Notice that the line for water would go through (0, 0) if you continued it in that direction. What does this mean in terms of density? How much weight do you have when you have no volume? How much volume do you have when you have no mass?

(f) The same is true for all objects made of matter. Each of your dozen objects seems to have only one point on the graph. If that were true, then we couldn't draw a line for its density. But, knowing that every material has no mass when it has no volume (and vice versa), we can use (0, 0) as the other point to draw each line. Draw the lines for each of the other 11 objects.

(g) Suppose that the point called *K* is ether. Which of the other 11 objects will float in ether? Which will sink?

(h) If object J is water from the Dead Sea, which of the other 11 objects will float in this? Which will sink in the Dead Sea?

(i) Suppose that object F is concrete. Could you build a concrete boat that would float in water? If yes, explain how. If no, explain why not.

Lynda Richardson
Informing the Public

Lynda Richardson is a journalist on the staff of *The New York Times*. She has won awards for her writing on special issues like AIDS and its spread.

"Each day, I have to be on the watch for new developments of the disease," Lynda states. "I also need to understand and report on the data that is released from many different scientific studies."

Lynda often uses graphs to show this data in visual form. "For example, with a graph we can clearly see how the rates of HIV infection have changed among various groups over time. Graphs also help me to track annual budget changes for AIDS research. I can then correlate these numbers to forecast the future of the disease."

Lynda's writing career began when she was a junior high school student in Tyler, Texas. "I was asked by an English teacher to write an essay about my life as an African American girl growing up in a small Texas town."

A longer version of this essay was later published by *The Dallas Times Herald*. Lynda went on to study journalism at the University of Texas. After graduating, she worked for several newspapers before landing the prestigious position at *The New York Times*.

"At my newspaper," she says, "I can have tremendous impact. Nothing gives me greater joy than to know that when I write a story it will have a positive impact on people's lives."

Graphical Estimation

CHAPTER
4

4.1 Graphs Versus Tables

Shana is celebrating her 12th birthday. She is 56 inches tall. How tall would you estimate she will be when she celebrates her 17th birthday? Explain how you came up with your estimate.

Estimating unknown quantities can be very difficult. In this chapter we look at some tools that are available for making estimates.

In Chapter 3 you used a rectangular coordinate system to plot straight lines. A rectangular coordinate system enables you to plot data sets that involve two variables. For example, consider Display 4.1, which shows a small data set involving age and height measurements for five women. The heights have been rounded to the nearest quarter inch, and ages have been rounded to the nearest year.

Age (in Years)	Height (in Inches)
12	57.5
14	60.25
17	63.75
14	62
13	56.25

Display 4.1

From the data set we form the data points

(12, 57.5) (14, 60.25) (17, 63.75) (14, 62) (13, 56.25)

We now plot these points as shown in Display 4.2.

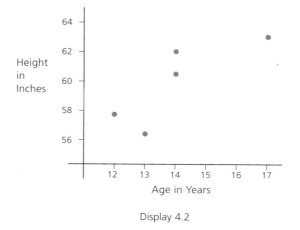

Display 4.2

A graph of data points such as the one shown in Display 4.2 is called a **scatterplot** or a **scattergram**.

Note: As Display 4.2 shows, when graphs are used to represent data from real world situations.

- **the unit of measure on the vertical axis may be different from the unit of measure on the horizontal axis; and**

- **the intersection of the vertical axis and the horizontal axis may not represent the point (0, 0).**

The more extensive scatterplot in Display 4.3 shows the age-height measurements of 45 women. All measurements were rounded to one decimal place.

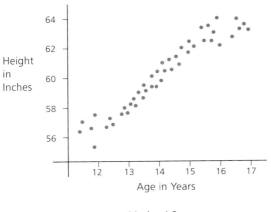

Display 4.3

Estimate the coordinates of the point x in Display 4.4. Describe in words the physical meaning of each coordinate you estimated.

a

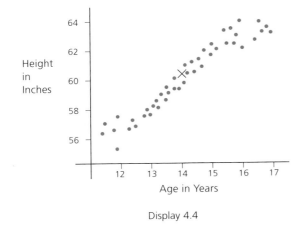

Display 4.4

The pattern of the data in Display 4.4 suggests something that you probably would expect—as age increases, so does height. List at least two other ideas suggested by this graph.

b

Shana is celebrating her 12th birthday. She is 56 inches tall. How tall would you estimate she will be when she celebrates her 17th birthday? Explain how you came up with your estimate.

c

Instead of writing a description of the variable on each axis, people often represent the variables and the axes by letters that abbreviate the intended meanings. For example, using A for age and H for height, the graph of Display 4.4 might appear with its axes labeled, as in Display 4.5.

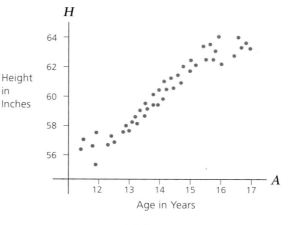

Display 4.5

Joseph, a student at Newton School, looked at the graph above and asked the teacher how the vertical and horizontal axes of a graph can intersect at a point other than (0, 0). The teacher said, "It's convenient, does not waste space, and makes a neater looking graph. It's the *in* thing!" Joseph still does not understand. How would you answer Joseph's question?

In this chapter we shall see how graphs of data help make decisions, estimate unknown quantities, answer questions, and show patterns. Because hospitals measure and record lots of data, let's look at an example in this area.

Your heart is a principal engineer of your life. With one hand make a fist. That is about the size of your heart.

It is the heart that pumps blood through the body at about 4 quarts per minute. Blood flow is essential for human life.

1. Using your age, approximately how many gallons of blood has your heart pumped through your body?

2. When you are 45 years old, approximately how many gallons of blood will your heart have pumped through your body?

One of the first measurements made when a person visits a doctor or enters the hospital is *blood pressure*. There are two types of blood pressure. One type measures the smallest pressure needed to stop the blood flow through the body. This pressure is called **systolic blood pressure**. The second type measures the largest pressure needed to resume the blood flow through the body. This type is called **diastolic blood pressure**. When a doctor gives a report of blood pressure as 125/85, this means that the systolic blood pressure is 125 while the diastolic blood pressure is about 85.

Suppose that a doctor has recorded the two types of blood pressure for 20 patients. The results are given in Display 4.6

Patient	Systolic Blood Pressure (SBP)	Diastolic Blood Pressure
1	160	90
2	150	75
3	135	85
4	110	70
5	130	75
6	145	60
7	120	80
8	130	80
9	145	85
10	120	70
11	100	50
12	90	50
13	110	65
14	90	55
15	135	60
16	160	100
17	100	65
18	110	75
19	125	70
20	145	70

Display 4.6

The average systolic blood pressure for healthy adults is about 120. The average diastolic blood pressure is about 75. These numbers will vary depending on whether a person is resting or involved in a physical activity.

Use Display 4.6 to answer these questions:

1. In terms of systolic blood pressure, would you say that a majority of these patients are low, normal, or high, as compared to healthy adults?

2. In terms of diastolic blood pressure, would you say that a majority of these patients are low, normal, or high, as compared to healthy adults?

Display 4.7 shows a scattergram of this data set. The horizontal axis indicates the systolic blood pressure while the vertical axis indicates the diastolic blood pressure. The point marked by an ⊗ indicates approximate average values for healthy adults.

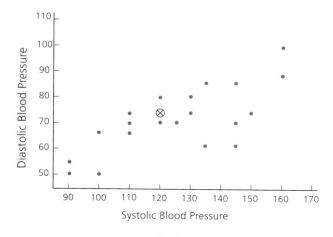

Display 4.7

Use Display 4.7 to answer these questions:

1. In terms of systolic blood pressure, would you say that a majority of these patients are low, normal, or high, as compared to healthy adults?

2. In terms of diastolic blood pressure, would you say that a majority of these patients are low, normal, or high, as compared to healthy adults?

Write a paragraph on whether you found it easier to use the table or the scattergram in answering questions about these patients. Explain.

The census of the United States is taken every 10 years. The census is an attempt to count the number of people in the U.S. Display 4.8 shows the census totals for the years 1790 to 2000. Display 4.9 is a graph of the same data.

1. Write two questions that you think are interesting and that would be easier to answer using the table rather than the graph.

2. Write two questions that you think are interesting and that would be easier to answer using the graph rather than the table.

To summarize, data can be presented in different ways. Using tables or graphs are the two most common methods. When there are few data points, a table may be the best way to look at the data. As the number of data points increases, getting a picture of "what's going on" from a table becomes difficult and it's time to graph the data. Usually one can read exact information from a table, while a graph provides reasonable estimates and an overall picture of the data.

Year	Population (in Millions)
1790	3.929
1800	5.308
1810	7.240
1820	9.638
1830	12.866
1840	17.069
1850	23.192
1860	31.443
1870	38.558
1880	50.156
1890	62.948
1900	75.995
1910	91.972
1920	105.711
1930	122.775
1940	131.669
1950	150.697
1960	179.323
1970	203.185
1980	227.224
1990	249.464
2000	281.421

U.S. Census

Display 4.8

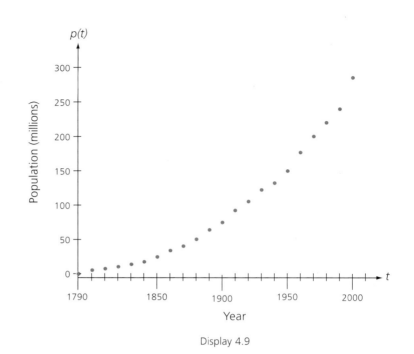

Display 4.9

When world conditions permit, Olympic track and field competitions are held every four years. Display 4.10 and Display 4.11 are a table and a graph of the winning times in seconds for the men's Olympic 800-meter race.

1. Why are there no Olympic records for the years between 1912 and 1920? Why are there no records for the years between 1936 and 1948?

2. Write a question that you think is interesting and that would be easier to answer using the table rather than the graph.

3. Write a question that you think is interesting and that would be easier to answer using the graph rather than the table.

Year	Winning Time (in Seconds)
1904	116.0
1908	112.8
1912	111.9
1920	113.4
1924	112.4
1928	111.8
1932	109.8
1936	112.9
1948	109.2
1952	109.2
1956	107.7
1960	106.3
1964	105.1
1968	104.3
1972	105.9
1976	103.5
1980	105.4
1984	103.0
1988	103.45
1992	103.66
1996	102.58
2000	105.08
2004	104.45

Years and Winning Times for
Men's Olympic 800-meter Race

Display 4.10

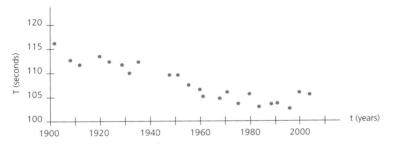

Years and Winning Times for Men's Olympic 800-meter Race

Display 4.11

Problem Set: 4.1

1. The number of insurance policies written by the Homeowners Mutual Insurance Company from 1997 to 2004 is given in Display 4.12. Make a scattergram of the data. Use the horizontal axis to represent the year and the vertical axis to represent the number of policies written. Do any of the points seem to be out of line with the rest? If so, explain why you think this happened.

Year	Number of Policies Written
1997	67,022
1998	73,724
1999	79,622
2000	81,215
2001	77,966
2002	80,123
2003	91,766
2004	93,601

Display 4.12

2. Display 4.13 gives the average speed in miles per hour of the winning auto in the Indianapolis 500 automobile race every other year from 1911 through 2003. Display 4.14 shows a graph of the same data but charts the race every 10 years.

(a) How do you think the speeds in the table were calculated?

(b) We forgot to graph the data for the year 1979, in which Rick Mears was the winner with an average speed of 158.9 miles per hour. Plot the point corresponding to this data.

(c) Why do you suppose there are gaps in the data corresponding to the years 1917, 1943, and 1945?

(d) What is the overall trend you observe? Is it easier to observe this trend from the table or from the graph?

(e) Do you notice any exceptions to this trend? Can you think of anything that might explain these exceptions?

(f) Write a question that you find interesting and that would be easier to answer from the table rather than from the graph.

(g) Write a question that you find interesting and that would be easier to answer from the graph rather than from the table.

Year	Winner's Avg. Speed (in mph)
1911	74.60
1913	75.93
1915	89.84
1917	
1919	88.05
1921	89.62
1923	90.95
1925	101.13
1927	97.55
1929	97.59
1931	96.63
1933	104.16
1935	106.24
1937	113.58
1939	115.04
1941	115.12
1943	
1945	
1947	116.34
1949	121.33
1951	126.24
1953	128.74
1955	128.21
1957	135.60
1959	135.86
1961	139.13
1963	143.14
1965	150.69
1967	151.21
1969	156.87
1971	157.74
1973	159.04
1975	149.21
1977	161.33
1979	158.90
1981	139.08
1983	162.12
1985	152.98
1987	162.18
1989	167.58
1991	176.46
1993	157.21
1995	153.62
1997	145.83
1999	153.18
2001	141.57
2003	156.29

Winner's Average Speed at the Indy 500 by Year[1]

Display 4.13

[1] From the Indianapolis 500 website, www.indy500.com.

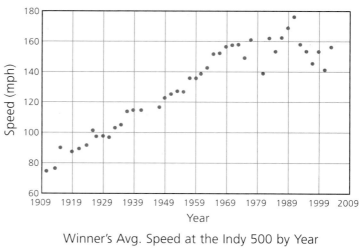

Winner's Avg. Speed at the Indy 500 by Year

Display 4.14

3. Display 4.15 lists the mean 2004 math SAT score for each state and the District of Columbia as reported by the *College Board*. Also in each listing is the percent of high school graduates who took the test.

(a) Using the scattergram of this data provided by your teacher, label each dot with the two letter postal code for the state.

(b) Examine your plot for trends. The states should appear to form two clusters (groups). Describe the features of each group.

(c) In general, as the percentage of students taking the SAT increases, what happens to the mean math score?

(d) Since there appear to be two distinct clusters, how would you answer the previous question for each cluster?

(e) Tiffany, who lives in Iowa, claims that students from Iowa are the best math students in the country, based on their SAT scores. Do you agree with her? Why or why not? Explain.

(f) How does your state compare with the other states? Do you think your state is doing well or poorly? Explain.

(g) Do you notice any differences among the various regions of the country? Explain.

State	% Participation Rate 2004	Mean SAT I Math
Alabama (AL)	10	553
Alaska (AK)	53	514
Arizona (AZ)	32	524
Arkansas (AR)	6	555
California (CA)	49	519
Colorado (CO)	27	553
Connecticut (CT)	85	515
D.C., Washington (DC)	77	476
Delaware (DE)	73	499
Florida (FL)	67	499
Georgia (GA)	73	493
Hawaii (HI)	60	514
Idaho (ID)	20	539
Illinois (IL)	10	597
Indiana (IN)	64	506
Iowa (IA)	5	602
Kansas (KS)	9	585
Kentucky (KY)	12	557
Louisiana (LA)	8	561
Maine (ME)	76	501
Maryland (MD)	68	515
Massachusetts (MA)	85	523
Michigan (MI)	11	573
Minnesota (MN)	10	593
Mississippi (MS)	5	547
Missouri (MO)	8	585
Montana (MT)	29	539
Nebraska (NE)	8	576
Nevada (NV)	40	514
New Hampshire (NH)	80	521
New Jersey (NJ)	83	514
New Mexico (NM)	14	543
New York (NY)	87	510
North Carolina (NC)	70	507
North Dakota (ND)	5	601
Ohio (OH)	28	542
Oklahoma (OK)	7	566
Oregon (OR)	56	528
Pennsylvania (PA)	74	502
Rhode Island (RI)	72	502
South Carolina (SC)	62	495
South Dakota (SD)	5	597
Tennessee (TN)	16	557
Texas (TX)	52	499
Utah (UT)	7	556
Vermont (VT)	66	512
Virginia (VA)	71	509
Washington (WA)	52	531
West Virginia (WV)	19	514
Wisconsin (WI)	7	596
Wyoming (WY)	12	546

2004 Math SAT Scores

Display 4.15

4.2 Linear Interpolation: Inserting Points Between Points

Learning Outcomes

After studying this section, you will be able to:

Explain the concept of linear interpolation

Use linear interpolation to estimate one of the variables between two known data points

Produce an equation for the line segment that contains two known data points as endpoints.

Photography is regarded as the world's most popular and widespread hobby. People take pictures of their families, of their vacation fun, of special times in their lives (like being promoted into high school), etc. Professionals take pictures for magazines and newspapers. Movies that we see at a theater are still made on film (not on videotape). Many, if not most, of you will take pictures some time in your lifetime. Some of you will become advanced amateurs or professionals and be able to use a darkroom for developing pictures.

In the darkroom one uses chemicals, normally mixed with water. Let's look at just one chemical that is used—the *developer*. In the darkroom, film is put into a "tank," a small tank looking much like a cup with a removable top. Developer is poured into the tank to produce images on the film from which pictures are eventually made.

When the developer is put into the tank, two very important measurements must be considered. One is the temperature of the developer; the other is the time that the film is allowed to remain in the developer. These items—temperature and time—are related. The higher the temperature, the less time should be allowed. In order to obtain good pictures, however, one must be precise. Indeed, with color film, one needs to be very precise.

Display 4.16 was enclosed with a roll of Kodak® film purchased at a local photography store. This table contains some very important information. The names and symbols in the left column refer to different types of developers. For example, T-MAX and D-76 are two different developers. The numbers across the top indicate possible temperatures of the developer, while the numbers in the table refer to minutes of developing time. For example, if D-76 developer is at 68°F, then, in a small tank, the film should be developed for 8 minutes.

KODAK T-MAX 400 Professional Film / 5053, 6053 (Rolls)

KODAK Developer or Developer and Replenisher	Developing Time (Minutes)									
	Small Tank (Agitation at 30-second intervals)					Large Tank (Agitation at 1-minute intervals)				
	65°F (18°C)	68°F (20°C)	70°F (21°C)	72°F (22°C)	75°F (24°C)	65°F (18°C)	68°F (20°C)	70°F (21°C)	72°F (22°C)	75°F (24°C)
T-MAX	NR	7	6½	6½	6	NR	7	6½	6½	6
T-MAX RS	NR	7	6	6	5	NR	8½	8	7½	7
D-76	9	8	7	6½	5½	10	9	8	7½	6½
D-76 (1:1)	14½	12½	11	10	9	—	—	—	—	—
HC-110 (Dil B)	6½	6	5½	5	4½	8	7	6½	6	5
MICRODOL-X	12	10½	9	8½	7½	13	11½	10	9	8
MICRODOL-X (1:3)	NR	NR	20	18½	16	—	—	—	—	—

Primary recommendations are in **bold** type.
NR = Not recommended

From KODAK T-MAX 400 Professional Film developing information insert.
©Eastman Kodak Company. Reprinted with permission.

Display 4.16

Using Display 4.16, decide which development time you would recommend for each of the following situations. Assume that all developers will be used in a small tank. Be prepared to explain how you got your answer.

1. D-76 developer at 75°F

2. D-76 developer at 71°F

3. D-76 developer at 63°F

4. T-MAX developer at 74°F

5. Microdol-X developer at 78°F

There are many different ways of attacking problems such as the ones above. One way is called *linear interpolation.* If we consider the part of the table that refers to D-76 in a small tank, we can write this information as follows:

Temperature (°F)	Time (Minutes)
F	T
65	9
68	8
70	7
72	6.5
75	5.5

Let's plot these points on a graph. There are two variables, F and T. We'll use the horizontal axis to represent the temperature F and the vertical axis to represent the time T. Display 4.17 shows the resulting graph.

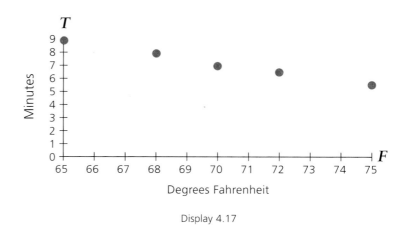

Display 4.17

Michael says, "These points lie on a straight line." Is Michael correct? Explain.

What do we do if the D-76 developer is at 67°F? There is no point on the graph where $F = 67$. Because 67 is between 65 and 68, let's zoom in and enlarge the portion of the graph that contains the points (65, 9) and (68, 8).

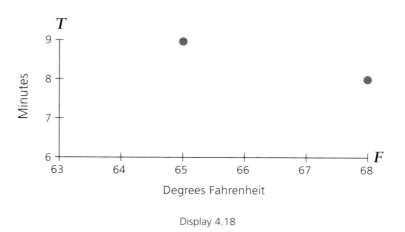

Display 4.18

If we connect the points (65, 9) and (68, 8) with a straight line segment, we get the picture shown in Display 4.19.

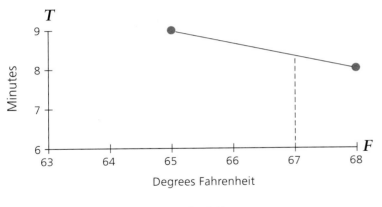

Display 4.19

1. There is a point on this line segment with an F-coordinate of 67. Estimate the T-coordinate of that point. In terms of minutes and seconds, what development time do you recommend?

a

2. While we were solving the problem above, the temperature of the D-76 developer rose to 67.5°. What development time do you recommend now?

We have called this method *linear interpolation.* According to the dictionary, *linear* means "related to a line," while *interpolate* means "to insert between two things." We used a line to insert a point between two points.

In order to find a T-coordinate for F = 67, we looked at the line segment joining (65, 9) and (68, 8). Scientists working for a film manufacturer are interested in the T-coordinate for many different values of F between 65° and 68°. In this case, estimating T becomes easier if one uses a formula. Such a formula comes from the equation of the straight line through (65, 9) and (68, 8).

Show that the equation of the line segment through (65, 9) and (68, 8) is given by

b

$$T = -\frac{1}{3}F + \frac{92}{3}$$

What is the value of T if F = 67? Does this answer agree with your previous estimate?

Note that the formula

$$T = -\frac{1}{3}F + \frac{92}{3}$$

was obtained only for values of F between 65 and 68, inclusive.

With an application of linear interpolation in mind, find formulas good for

1. values of F between 68 and 70, inclusive

2. values of F between 70 and 72, inclusive.

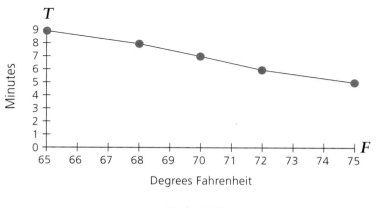

Display 4.20

Geometrically, we are using the graph shown in Display 4.20, which consists of several straight line segments, to find estimates of development times for any temperature between 65° and 75°.

You are working for the film processing department of the Ritzo Camera Store. It's a hot summer day, and your boss tells you that the D-76 developer is above 75°. She insists that you extend the graph in Display 4.20 to include temperatures between 75 and 80, inclusive. Your job depends on it. How might you go about doing this? Remember, it's your responsibility to make good pictures.

Quick Photo uses Microdol-X in a small tank to develop films. Kenji finds the developer temperature is 67°F. He needs to use linear interpolation to estimate the developing time. Kenji asks you to assist by

1. choosing the appropriate interval and drawing the graph

2. finding the equation of the appropriate line segment

3. using the equation to estimate the developing time

4. indicating the appropriate point on the graph.

   Kenji now asks you to write the directions for finding the Microdol-X developing time for a temperature of 74°F.

Now let's use linear interpolation with the scattergram involving diastolic blood pressure which we discussed in Section 4.1. It is reprinted here, as Display 4.21, for your convenience.

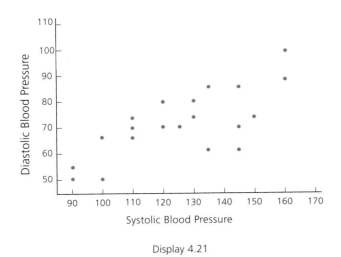

Display 4.21

There is no point on this graph with a systolic blood pressure reading of 115. How might you find such a point by using linear interpolation? Would you regard such a point as a good estimate? Explain.

Now let's return to the population displays for the United States during the years 1790 to 2000. They are shown again in Display 4.22 and Display 4.23.

Year	Population (in Millions)
1790	3.929
1800	5.308
1810	7.240
1820	9.638
1830	12.866
1840	17.069
1850	23.192
1860	31.443
1870	38.558
1880	50.156
1890	62.948
1900	75.995
1910	91.972
1920	105.711
1930	122.775
1940	131.669
1950	150.697
1960	179.323
1970	203.185
1980	227.224
1990	249.464
2000	281.421

Display 4.22

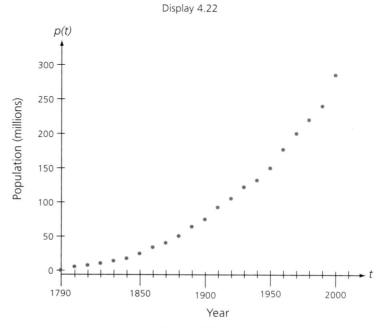

Display 4.23

From the census figures of 1840 and 1880, find the equation of a straight line for estimating the population at any time between 1840 and 1880. Use your formula to estimate the population in 1850, 1860, and 1870. Are your estimates of the population by linear interpolation larger or smaller than the actual census figures? Could you have predicted this by looking at the graph? Explain.

a

Find formulas for estimating the population any time between (i) 1840 and 1850, (ii) 1850 and 1860, (iii) 1860 and 1870, and (iv) 1870 and 1880. For each of the line segments obtained in this way, list the slopes in your own copy of this table

b

Years	Slope
1840–1850	
1850–1860	
1860–1870	
1870–1880	

Compare the slopes and write a summary of your conclusions. What happened in the United States during the 1860s? Is there any relation between your answer to this question and the slopes you computed?

Problem Set: 4.2

Problems 1 and 2 refer to Display 4.24, which shows the number of Democratic U.S. Senators in Congress during some of the years from 1953 to 2005.

1. Would the table be more helpful if it indicated the number of Democratic Senators *each year*, rather than *every two years*? Explain.

Congress	Senate Democrats
1953	47
1955	48
1957	49
1959	65
1961	64
1963	66
1965	68
1967	
1969	
1971	54
1973	56
1975	60
1977	61
1979	58
1981	
1983	46
1985	47
1987	55
1989	55
1991	56
1993	57
1995	48
1997	45
1999	45
2001	50
2003	48
2005	44

Number of Senate Democrats

Display 4.24

2. Suppose you researched this data for an assignment due today and, just as you are about to hand it in, you realize that you forgot to record the data for 1967, 1969, and 1981. Use linear interpolation to estimate the number of Senators for those years and plot these missing points on a copy of the graph shown in Display 4.25.

Then:

(a) Go to your library and find the correct answers. One source is the *World Almanac*.

(b) Comment on how closely you estimated the correct answers.

(c) One of your estimates is probably better than the other. Which is it, and why?

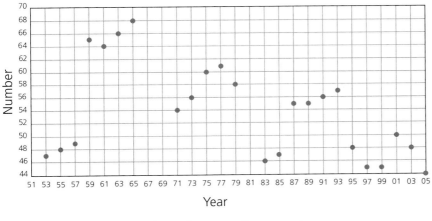

Year

Number of Senate Democrats

Display 4.25

3. Display 4.26 is the same one you used in the problem set of Section 4.1. It gives the average speed in miles per hour of the winning auto in the Indianapolis 500 automobile race every other year from 1911 through 2003. Although the table contains entries for the odd years, the race is run every year.

(a) Use linear interpolation to estimate the winning speed for each of the years 1926, 1974, and 1990.

(b) Ask your teacher for the official recorded speeds for those years. Do you notice any differences between the official speeds and your estimates? What factors do you think might have caused the differences? An almanac or other sports data source might help you.

Year	Winner's Avg. Speed (in mph)
1911	74.60
1913	75.93
1915	89.84
1917	
1919	88.05
1921	89.62
1923	90.95
1925	101.13
1927	97.55
1929	97.59
1931	96.63
1933	104.16
1935	106.24
1937	113.58
1939	115.04
1941	115.12
1943	
1945	
1947	116.34
1949	121.33
1951	126.24
1953	128.74
1955	128.21
1957	135.60
1959	135.86
1961	139.13
1963	143.14
1965	150.69
1967	151.21
1969	156.87
1971	157.74
1973	159.04
1975	149.21
1977	161.33
1979	158.90
1981	139.08
1983	162.12
1985	152.98
1987	162.18
1989	167.58
1991	176.46
1993	157.21
1995	153.62
1997	145.83
1999	153.18
2001	141.57
2003	156.29

Winner's Average Speed at the Indy 500 by Year

Display 4.26

4.3 Least-Squares Lines: Lines Replacing Points

Display 4.27 and Display 4.28 show the actual monthly sales data for a company that makes x-ray film for hospitals, doctors, etc. We'll call the company the Codac Film Co. The sales numbers, in terms of x-ray film sheets sold, should be given for each month of a one-year period, beginning in January (month 1).

Learning Outcomes

After studying this section, you will be able to:

Describe lines that come close to representing a data set

Explain the meaning of a least-squares line

Use the least-squares line to make estimates from data points.

Month	Film Sheets Sold
1	21,000
2	23,000
3	41,000
4	32,000
5	28,000
6	41,000
7	
8	40,000
9	47,500
10	48,000
11	51,000
12	43,000

X-ray Film

Display 4.27

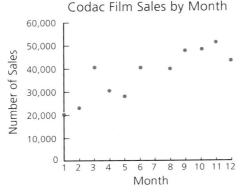

Codac Film Sales by Month

Display 4.28

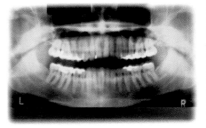

243

Notice that there is no point on the graph for month 7. The reason is the following: During July, the company hired several students from Newton School. Because of their inexperience, they accidentally destroyed all sales records for July. The IRS wants taxes on those sales. (Do you know what *IRS* stands for?) An examiner from the IRS indicates she is willing to accept a reasonable estimate for the July sales. The president of Codac has called the staff together to come up with an estimate of the number of sales in July. Marcello, one of the staff members, suggests the method of linear interpolation using the number of sales in month 6 and month 8.

If Marcello uses linear interpolation by looking at the number of sales in month 6 and month 8, what estimate would Marcello make? Explain how you think Marcello would get his estimate.

Ellen, another staff member, disagrees with Marcello's method. She argues that linear interpolation does not pay any attention to all of the other sales numbers for the year. She says that the sales numbers during the year are widely scattered and it would be unfair to use only the number of sales in months 6 and 8. Ellen suggests that one straight line be drawn to represent all the data points, coming as close to them as possible. Then an estimate of the number of sales in July can be obtained by looking at the value on the vertical axis that corresponds to 7 on the horizontal axis.

Your teacher will give you a copy of the scattergram in Display 4.28. On your copy of the scattergram, use a ruler to draw a line that you think comes "close" to the data points. Now estimate the sales for July by looking at the value on the vertical axis corresponding to 7 on the horizontal axis.

1. **What is your estimate of the number of sales in July?**

2. **How does your estimate compare with that of your classmates?**

3. **How does your answer compare with Marcello's, which he got using linear interpolation?**

Ellen's method and Marcello's method are illustrated by the graphs in Display 4.29. Marcello's estimate of the number of sales for July (month 7) is 40,500, while Ellen's estimate is 37,000.

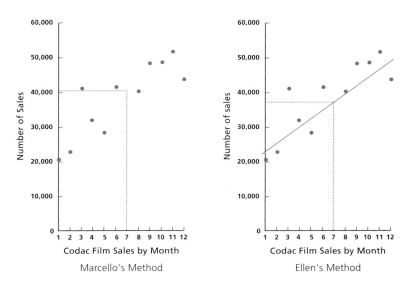

Display 4.29

a

Which of the two methods of estimating do you find more acceptable—Ellen's or Marcello's? Why? Which method do you think the president of Codac would prefer? Which method do you think the IRS would prefer? Explain your answers.

b

At Newton School, 30 students looked at the graph of Codac's sales data and drew lines that they felt came close to the data points. Many different lines were drawn. Your class has been asked to act as a judge and pick the best straight line. As a class, do you have ideas about what you are looking for? Write a summary that contains the ideas of your class.

Ellen's idea gives us another method for creating new points, a method usually called "eyeballing." There are some difficulties with "eyeballing." For example, in a class of 30 students, we might get 30 different straight lines and 30 different estimates. Another difficulty is this: Not only have we created new points, but if we are going to use a straight line to represent all the data points, then we have replaced the original data points by points on that straight line. In our example, the original data set indicated that during month 3, the number of sales was 41,000. If we use Ellen's straight line, then during month 3 it would appear that the number of sales is about 27,000. This second difficulty will be discussed later.

The first difficulty can be overcome if we agree on a specific way to draw our straight line to come close to the data points. Great mathematicians such as Laplace, Legendre, and Gauss considered various ways of finding a line to represent the data points in a scattergram. In order to demonstrate the ideas of these mathematicians, let's consider the data points $(1, 1)$, $(2, 3)$, $(3, 7)$, and $(4, 6)$ on this graph.

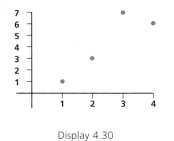

Display 4.30

In Display 4.31, a straight line has been drawn close to those data points, and the vertical distances d_1, d_2, d_3, and d_4 have been labeled. Historically, a number of people were interested in finding a straight line that made the sum of these distances (in this case, $d_1 + d_2 + d_3 + d_4$) as small as possible. Note that d_1 is the vertical distance from the first data point to the straight line; d_2 is the vertical distance from the second data point to the straight line, etc. Remember that a "distance" is never a negative number.

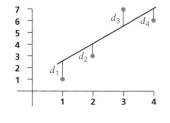

Display 4.31

1. Your teacher will give you a copy of a coordinate plane in which the unit of measure on the vertical axis is 1 centimeter. In this plane, sketch the points (1, 1), (2, 3), (3, 7), and (4, 6). Using a ruler, draw (eyeball) a straight line that comes "close" to the data points. Then draw vertical segments from each data point to your line. Now, using your ruler, find the sum

$$d_1 + d_2 + d_3 + d_4$$

to the nearest 10th of a centimeter. Make a listing of all the sums obtained by members of the class. Which is the smallest?

2. The mathematician Laplace was interested in finding a line for which the largest of the numbers d_1, d_2, d_3, and d_4 is as small as possible. Make a listing of the largest vertical distance obtained by each member of the class. Which one of these is the smallest?

Some historical comments at this point might help your understanding. In 1757, a mathematician named Boscovich wanted to find straight lines that made sums like

$$d_1 + d_2 + d_3 + d_4$$

as small as possible. He was unable to find a method. In 1786, the mathematician Laplace wanted to find straight lines that made the largest of the numbers d_1, d_2, d_3, and d_4 as small as possible. He, too, was unable to find a method. In 1805, Legendre published a method that made a sum of squares like

$$d_1^2 + d_2^2 + d_3^2 + d_4^2$$

as small as possible. In 1806, the mathematician Gauss claimed that he had discovered this method before Legendre. No one is sure what actually happened.

Using the same graph and vertical distances you found in the previous problem, find the sum

$$d_1^2 + d_2^2 + d_3^2 + d_4^2$$

to the nearest tenth of a square centimeter. Make a listing of all the sums obtained by members of your class. Which one is the smallest? Were you surprised?

The method that minimizes the sum of the squares of the vertical distances is known as the method of **least squares.** A great deal is known about the method of least squares, and thus it has become quite popular. Indeed, this method is used so often that calculators and computers have built-in programs to find straight lines using this idea. This is the topic of the next section. It should be clear, however, that there has been no universal agreement on what is meant by the "best" straight line to represent a data set.

Problem Set: 4.3

1. Consider the data points $(-1, 2)$, $(0, 3)$, $(1, 1)$, and $(2, 4)$. Draw a scattergram. A group of students working together is asked to find the equation of a straight line that comes close to these data. Mary gives the equation

$$y = 0.5x + 2.7$$

while Kim gives the equation

$$y = -x + 3$$

Your teacher will give you a graph of Mary's and Kim's lines and a chart to help answer these questions.

(a) Use a calculator to complete the chart.

(b) Now determine which line is better in the sense of least squares.

(c) Which line would Laplace regard as the better of the two?

(d) If you want $d_1 + d_2 + d_3 + d_4$ as small as possible, which of the two lines is better?

(e) See if you can find the equations of lines that are better than either Mary's or Kim's

(i) in the sense of least squares,

(ii) in the sense of pleasing Laplace, and

(iii) in the sense of keeping $d_1 + d_2 + d_3 + d_4$ small.

2. Display 4.32 lists the percentage of 12–17 year-old students who reported using illicit drugs at any time, as indicated by the National Household Survey on Drug Abuse.

Percentage of 12–17 Year-Olds Reporting They Have Used Illicit Drugs	
Year	Percent
1979	34.3
1982	27.6
1985	29.5
1988	24.7
1990	22.7
1991	20.1

Display 4.32

Display 4.33 shows a scattergram for the data in the table. Also graphed on the scattergram is the line that Amy claims "best" represents the data. The vertical distances d_1 through d_6 have been drawn.

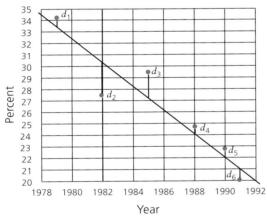

Percentage of 12–17 Year-Olds Reporting
They Have Used Illicit Drugs

Display 4.33

(a) Using your ruler, find the sum

$$d_1^2 + d_2^2 + d_3^2 + d_4^2 + d_5^2 + d_6^2$$

to the nearest millimeter.

(b) Make a list of all the sums computed by you and your classmates.

(c) Calculate the three measures of center for the sums in this list.

(d) Which one of these measures of center do you think best represents the data? Explain. This is an important number that you will need later. Let's give it a name: call it Z.

(e) On the copy of the scattergram provided by your teacher, use a ruler to draw a line that you think comes closer to the data points. Using your ruler, draw the vertical distances and find the sum

$$d_1{}^2 + d_2{}^2 + d_3{}^2 + d_4{}^2 + d_5{}^2 + d_6{}^2$$

to the nearest millimeter.

(f) Make a list of all the sums computed by you and your classmates. Is any one of them smaller than Z? Do you agree with Amy's claim?

4.4 Least-Squares Lines on a Graphing Calculator

The least-squares line (also called a **regression line**) is used frequently enough that the job of finding such a line is built into a number of calculators and pieces of computer software. Appendix A shows you how to obtain this line on a TI-82 (TI-83). If you are familiar with this procedure, move on, my fearless friend. Otherwise, go to the Appendix and check the steps required.

Learning Outcomes

After studying this section, you will be able to:

Use a calculator to determine the slope and y-intercept of a least-squares line

Use a calculator to make estimates from data points.

For example, if the data points (−1, 2), (0, 3), (0.5, 4), (1, 7), and (1.5, 3) are inserted, a graphing calculator will give you the line

$$y = 1.05x + 3.38$$

rounded to two decimal places.

Consider this two variable data set, where x represents the age of a child, in years, and y represents the height (or length), in centimeters. 1.7 represents 1 year 7 months; 0.6 represents 6 months. What does 3.8 represent?

x	y
1.7	72
3.2	88
3.1	85
0.6	63
1.2	77
1.8	84
3.8	100

Plot these points on a coordinate graph. Using a ruler, draw a line that you think comes close to these data. Now use your graphing calculator to find the least-squares line. How would you compare your line with the one from your graphing calculator?

Here is an actual application of the least-squares line in science:

Paleontologists have to deal with very old things. Almost every living thing gives away its age, but when you are looking at the remains of things that have been dead a long time, how can you tell their ages? Digging up fossils is a great deal of fun and hard work, and is extremely important for finding out how our planet works. There is no Owners Manual. Finding the exact age of a fossil is usually quite difficult. If you have no idea of the age of a fossil, there is a complicated way to find it. Every thing that has ever lived has a small, natural amount of radioactivity in it. While you are alive, this amount stays the same, but when you die, the radioactive chemical carbon 14 (or C-14) begins to decay, and its amount decreases in a predictable way.

To find out how old a fossil is, one extracts some of the radioactive chemical and measures it in an expensive piece of equipment to determine the percent of C-14 remaining. Then one uses the information in Display 4.34 to determine the age of the fossil.

1. A fossil has 7.15% of its C-14 left. Estimate its age.

2. If an animal died 40,000 years ago, what percentage of its C-14 would be left in its fossil?

3. If a fossil has 5.17% of its C-14 left, estimate the age of the fossil.

4. If a fossil has 50% of its C-14 left, use linear interpolation to estimate the age of the fossil. This number of years is known as the *half-life* of C-14.

The process we have just described and used is called *carbon dating*. It is expensive and takes a long time. Carbon dating is a process that scientists do not want to use very often.

It turns out that scientists don't have to use it very often. If you take samples up and down a canyon cliff, where there are layers of fossils, laid down one after the other, you can use a shortcut. When you dig or look at a cliff face of fossils, the topmost ones must have been put there last and they must be the youngest. It's hard to imagine that any might have slipped in out of order, down at the bottom!

Years Dead	C-14 Left
1	99.99%
1000	88.70%
2000	78.68%
3000	69.78%
4000	61.90%
5000	54.90%
6000	48.70%
7000	43.19%
8000	38.31%
9000	33.98%
10,000	30.14%
11,000	26.74%
12,000	23.72%
13,000	21.04%
14,000	18.66%
15,000	16.55%
16,000	14.68%
17,000	13.02%
18,000	11.55%
19,000	10.24%
20,000	9.09%
21,000	8.06%
22,000	7.15%
23,000	6.34%
24,000	5.62%
25,000	4.99%
26,000	4.42%
27,000	3.92%
28,000	3.48%
29,000	3.09%
30,000	2.74%
31,000	2.43%
32,000	2.15%
33,000	1.91%
34,000	1.70%
35,000	1.50%
36,000	1.33%
37,000	1.18%
38,000	1.05%
39,000	0.93%
40,000	0.83%
41,000	0.73%
42,000	0.65%
43,000	0.58%

Percent C-14 Remaining in a Fossil

Display 4.34

Consider a 400-foot cliff. Suppose that you can afford to pay for six C-14 tests. You have available sample fossils from the top to bottom at 2 feet, 89 feet, 120 feet, 210 feet, 297 feet, and 399 feet. Here are the results of the C-14 tests.

Depth (feet)	Time Since Death (years)
2	1466
89	66,038
120	102,840
210	154,980
297	273,240
399	301,245

One way of looking at these data would be to label a diagram of the cliff showing the number of feet from the top of the cliff and time since death, as in Display 4.35.

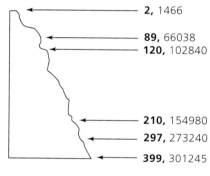

Display 4.35

We might also consider this information as a set of data points to graph. The first coordinate indicates the number of feet from the top of the cliff and the second coordinate indicates the number of years since death (Display 4.36).

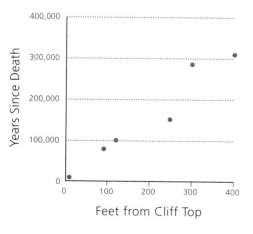

Display 4.36

The result is fairly predictable. As the number of feet from the cliff top increases, so do the fossil's years since death. This means we can use a simple method—record the fossil's depth from the cliff top as we find it—instead of a complicated method like carbon dating.

Using your graphing calculator, find the least-squares line for this set of data points.

a

The data and the least-squares line are plotted on the same graph in Display 4.37.

The least-squares line helps to fill in the points where we didn't find any samples, and helps us predict the age of samples found at previously unseen positions.

Using the least-squares line, estimate the age of fossils found at these distances from the top of the cliff:

b

(a) 132 ft.   **(b) 64 ft.**   **(c) 358 ft.**

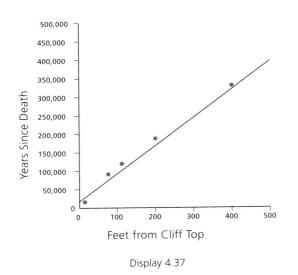

Display 4.37

If an animal died 230,000 years ago, estimate how far from the top of the cliff its fossil remains would be buried.

c

Why are fossils buried?

Carbon dating was a great scientific discovery. Who discovered it? When was it discovered? Find an article about a recent use of carbon dating and write a paragraph about it. (*Suggestion:* Go to your library to look for answers to these questions.)

Problem Set: 4.4

1. The ozone layer that exists high in the atmosphere is a major concern of scientists. This layer of ozone (oxygen atoms bound together in threes) filters out harmful ultraviolet light from the sun. The ozone layer has been monitored since the early 1970s. The layer is measured in units called Dobsons. Display 4.38 shows measurements recorded during the period between 1971 and 1985.

Year	Level (in Dobsons)
1971	300
1972	305
1979	260
1980	225
1981	245
1982	235
1983	205
1984	195
1985	180

Ozone Measurements 1971–1985

Display 4.38

There was a gap in the measurements between 1972 and 1979. Using a graphing calculator, find the least-squares line for this data set. Using this least-squares line, estimate the ozone levels for the missing years between 1972 and 1979.

2. The counselor at Newton School was concerned about students' scores on their final exam. He was able to collect the following data from some of the students in Ms. Stego's math class. Students reported the number of hours they studied for the exam, and then Ms. Stego recorded the exam score. Display 4.39 shows the data.

   (a) Make a scattergram for the data.

   (b) Susan, a student in the class, studied eight hours. If we use a graphing calculator to find the least-squares line that comes "close" to the data points, what estimate would you make on Susan's score?

   (c) Don wanted to use linear interpolation to estimate Susan's score. Comment on Don's idea.

Hours Studied	Exam Score
0	50
1	60
2	65
3	70
3	75
4	70
4	80
5	75
5	85
6	85
7	75
7	85
7	90
7	95
9	80
9	90
9	95
10	85
10	95
10	100

Ms. Stego's Math Class

Display 4.39

3. Display 4.40 lists the percentages of 12–17 year-old students who reported using illicit drugs at any time. We used this data in a previous problem. Display 4.41 shows a scattergram for these data. Also graphed is the line which Amy claims best represents the data.

   (a) Using your graphing calculator, find the least-squares line for the data in the table.

   (b) On the copy of the scattergram provided by your teacher, draw the line, using a ruler, that is represented on your graphing calculator screen. How does the line from your calculator compare with Amy's line?

Percentage of 12–17 Year-Olds Reporting They Have Used Illicit Drugs	
Year	Percent
1979	34.3
1982	27.6
1985	29.5
1988	24.7
1990	22.7
1991	20.1

Source: National Household Survey on Drug Abuse

Display 4.40

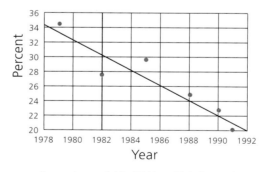

Percentage of 12–17 Year-Olds Reporting
They Have Used Illicit Drugs

Display 4.41

4.5 Forecasting

A grocery store in a small New England town obtains milk from a local dairy farmer. The store orders milk from the farmer on a weekly basis. It is important for the store manager to predict (estimate) how many gallons will be sold next week.

Why is it important for the store manager to be able to predict the number of gallons to be sold next week? List as many reasons as you can.

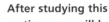

a

Here are the weekly milk sales for the last five weeks:

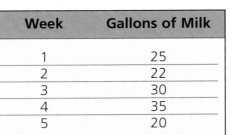

Week	Gallons of Milk
1	25
2	22
3	30
4	35
5	20

Predict how many gallons of milk will be sold next week (Week 6). Then write a paragraph defending your prediction.

b

The store manager's problem involves making a prediction of a future event based on previously gathered data. Making such predictions is known as *forecasting*. Actually, several problems you've already met have involved forecasting. Forecasting is not an exact science. A person who works in forecasting is known as a *forecaster*. Professional forecasters are usually in great demand and are able to find jobs with good salaries. In this section we examine a few of the methods used by forecasters.

Learning Outcomes

After studying this section, you will be able to:

Distinguish between interpolation and extrapolation

Use the least-squares line in forecasting

Explain the limitations of extrapolation

Describe patterns contained in data sets.

Before going on, there are a few terms that you need to understand.

A Word to Know: The process of making an estimate within the region of a given set of data points is called **interpolation** (as in *linear interpolation*).

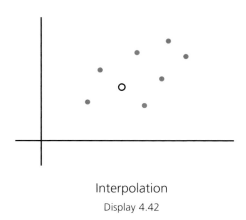

Interpolation
Display 4.42

A Phrase to Know: The process of making an estimate outside and to the left of the region of the data points is called **backward extrapolation**.

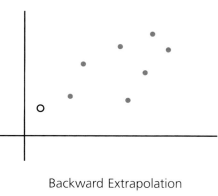

Backward Extrapolation
Display 4.43

A Phrase to Know: The process of making an estimate outside and to the right of the region of the data points is called **forward extrapolation** or **forecasting**.

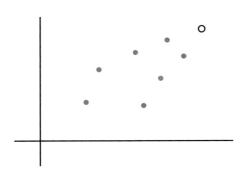

Forward Extrapolation—Forecasting

Display 4.44

The word **extrapolation** normally refers to either backward or forward extrapolation. The next example illustrates forecasting (forward extrapolation).

The 800-meter race is considered to be one of the more difficult races in the Olympic games because it requires an athlete to have the speed of a sprinter and the endurance of a distance runner. The fastest (winning) men's times (in seconds) for the Olympic 800-meter run have been collected from 1904 to 2000. This information appears in Display 4.45. A graph of the data is shown in Display 4.46.

If the Olympic games had been held in 1974, what estimate would you give on the fastest running time at those games? Be able to explain your answer. Does this problem involve interpolation or extrapolation? Is this a forecasting problem?

a

Based on the given information, what would you predict to be the fastest running time at the 2004 Olympic games? Be able to explain your answer. Does this problem involve interpolation or extrapolation? Is this a forecasting problem? What was the fastest running time at the 2004 Olympics?

b

Year	Winning Time (in Seconds)
1904	116.0
1908	112.8
1912	111.9
1920	113.4
1924	112.4
1928	111.8
1932	109.8
1936	112.9
1948	109.2
1952	109.2
1956	107.7
1960	106.3
1964	105.1
1968	104.3
1972	105.9
1976	103.5
1980	105.4
1984	103.0
1988	103.45
1992	103.66
1996	102.58
2000	105.08

Years and Winning Times for the
Men's Olympic 800-meter Race

Display 4.45

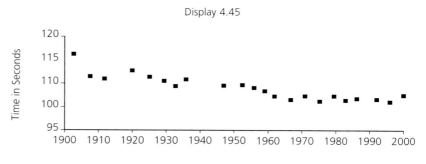

Years and Winning Times for the
Men's Olympic 800-meter Race

Display 4.46

A popular topic of discussion among track and field fans is

At which Olympics will the 800-meter race be run in less than 100 seconds?

Using any reasonable method you want, predict at which Olympic games the 800-meter race will be run in less than 100 seconds. Explain how you made your prediction.

a

- Using a graphing calculator or a spreadsheet, find the least-squares line through the Olympic data set. Use this straight line to predict in which Olympic games the 800-meter race will be run in less than 100 seconds.

- Now, instead of using all of the data points, use only those points where the year is 1960 or later. Again, using your graphing calculator or a spreadsheet, find the least-squares line through this data set. Use this straight line to predict in which Olympic games the 800-meter race will be run in less than 100 seconds.

Write a paragraph comparing your two predictions and discussing which one you believe to be more accurate.

b

1. The Funky Sneakers Shoe Co. has just announced the design of a space age running shoe that will reduce the average 800-meter running time by three seconds. At which Olympic game do you now predict the 800-meter race will be run in less than 100 seconds?

2. Besides a new type of running shoe, what other influences could affect Olympic records?

Susan drew a straight line through the Olympic data set as shown. It is her prediction that the 800-meter race will be run in less than 100 seconds at the 2004 or 2008 Olympics.

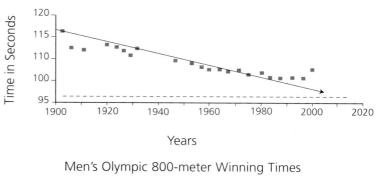

Men's Olympic 800-meter Winning Times
Display 4.47

By including more numbers on the vertical and horizontal axes, Susan was able to predict what might happen way in the future (when *Star Wars* will be raging).

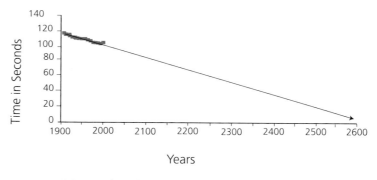

Men's Olympic 800-meter Winning Times
Display 4.48

According to Susan's graph above, in the year 2600, athletes will be running the 800-meter race in 0 seconds!

Write a paragraph on Susan's prediction that athletes will be running the 800-meter race in 0 seconds in the year 2600.

In the previous problems we have been looking only at the men's records in the 800-meter race. Do you think that a woman will run the 800-meter race in 100 seconds (or less) before a man does? [*Hint*: Look in an almanac (go to the library if necessary) for the women's track and field records from the Olympics. Now use your ideas on forecasting.]

We began this section by looking at a small grocery store and the problems the manager faced in terms of predicting milk sales. We looked at only five weeks of previously gathered data. Previously gathered data is called history by forecasters. Many times there is a pattern or set of trends that is apparent only when one looks at a good deal of history. For example, our five week data set was

Week	Gallons of Milk
1	25
2	22
3	30
4	35
5	20

It is highly unlikely that anyone would see a pattern or a trend in this data set. However, if we look at the 16 weeks of history given in Display 4.50 and sketch a graph of the data as shown below, then a trend does appear.

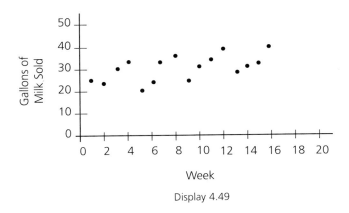

Display 4.49

Week	Gallons of Milk
1	25
2	22
3	30
4	35
5	20
6	25
7	34
8	38
9	23
10	28
11	32
12	37
13	26
14	29
15	31
16	40

Grocery Store Milk Sales

Display 4.50

Do you see it? The milk sales display a four week repeating pattern with the suggestion of an increasing trend.

 Forecast the milk sales for weeks 17, 18, 19, and 20. Be prepared to defend your answers.

REFLECT

In this chapter you have been introduced to formal methods for making estimates. The art or science of estimation is important in conducting our daily lives as well as an invaluable tool for being productive in many jobs. You should have the idea that graphs form a powerful tool for getting an overview of data sets, while tables are frequently useful in obtaining more precise information. The ideas of estimation, whether specifically stated or not, will permeate much of the mathematics that you see in later chapters.

Problem Set: 4.5

1. The number of insurance policies written by Homeowners Mutual Insurance Company each year from 1991 to 2001 are shown in Display 4.51. Forecast the number of policies to be written in 2003 and 2004. Explain how you arrived at your answers.

Year	Number of Policies Written
1991	53,000
1992	59,820
1993	
1994	
1995	67,022
1996	73,724
1997	79,622
1998	81,215
1999	77,966
2000	80,123
2001	91,766

Insurance Policies Written

Display 4.51

2. Owners of the Berlyn Cinema movie theater wanted to know how movie attendance varied by age, so they conducted a poll. Display 4.52 shows the results for 16 of the people polled.

   (a) Draw a scatterplot for the data. You will have to decide what scale is best.

   (b) Use a graphing calculator to determine the least-squares line for this data set and record the formula for the line. Draw the line on a scatterplot.

   (c) Using the formula from part (b), estimate how many movies a 15 year-old moviegoer would attend in a year. Discuss whether or not the estimate is reasonable. Include arguments for and against your estimate and the estimates of your classmates.

Age	Movie Per Year
16	30
17	28
17	24
18	24
18	26
19	20
20	16
21	12
22	12
23	20
24	12
24	16
25	8
25	12
26	8
26	10

Annual Movie Attendance by Age

Display 4.52

3. The following list of managers' salaries at Shapiro's Department Store was compiled at the end of the year.

Experience	Salary ($)
1 year	28,000
11 years	42,000
5 years	34,000
6 years	34,000
7 years	40,000
12 years	45,000
6 years	26,000
10 years	42,000

Make a scatterplot of the information, find the least-squares line, draw that line on the scatterplot, and record the formula for the line. Then answer these questions:

(a) What would be the starting salary for a new manager with no experience? Use backward extrapolation to make an estimate.

(b) Next year a new manager will be hired. That person will have 20 years experience. Using your least-squares line again, forecast what salary should be offered to this new manager. Do you think the owners of the department store will want to pay this salary? Why or why not?

4. The following data are the actual monthly sales numbers, over a period of 35 months, for a company that makes x-ray film for hospitals, doctors, etc. We used part of this data in a previous example. We called this company the Codac Film Company. The data are given in the table below and also as a scattergram in Display 4.54. Your job is to forecast sales for the next three months—36, 37, and 38.

Month	Film Sheets Sold	Month	Film Sheets Sold
1	21,000	19	35,416
2	23,000	20	34,885
3	41,000	21	41,832
4	32,000	22	38,515
5	28,000	23	33,448
6	41,000	24	33,370
7	42,000	25	38,428
8	40,000	26	31,160
9	47,500	27	33,798
10	48,000	28	36,379
11	51,000	29	51,374
12	43,000	30	20,544
13	29,500	31	41,552
14	32,000	32	16,767
15	50,000	33	024,694
16	29,000	34	32,839
17	40,000	35	15,172
18	49,935		

Display 4.53

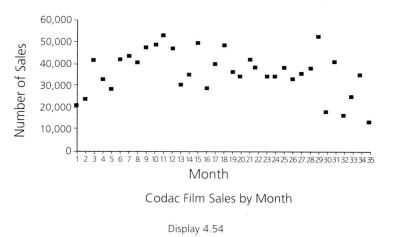

Codac Film Sales by Month

Display 4.54

5. Display 4.55 shows the winning times in seconds for the men's and women's Olympic 400-meter freestyle swimming competition. The 2000 and 2004 times are missing.

(a) Use a calculator or a spreadsheet to find an equation for a least-squares line for the women's times. Use your equation to predict the winning time for the 2000 women's Olympic 400-meter freestyle competition. Then go to the library and look up the actual time. Using the 2000 time, predict the time for the 2004. What was the fastest time at the 2004 Olympics?

(b) Use a calculator or a spreadsheet to find an equation for a least-squares line for the men's times. Use your equation to predict the winning time for the 2000 men's Olympic 400-meter freestyle competition. Then go to the library and look up the actual time. Using the 2000 time, predict the time for the 2004. What was the actual fastest time at the 2004 Olympics?

(c) Plot the graphs for both lines on a graphing calculator to decide if the women's time will ever be faster than the men's. When, if ever, do you predict that this will occur?

Year	Women's Time (in Seconds)	Men's Time (in Seconds)
1924	362.2	304.2
1928	342.8	301.6
1932	328.5	288.4
1936	326.4	284.5
1948	317.8	281
1952	312.1	270.7
1956	294.6	267.3
1960	290.6	258.3
1964	283.3	252.2
1968	271.8	249
1972	259.04	240.27
1976	249.89	231.93
1980	248.76	231.31
1984	247.1	231.23
1988	243.85	226.95
1992	247.18	225.00
1996	247.25	227.97

Olympic 400-m Freestyle

Display 4.55

PROJECT

The Gaitur Aid Company makes a leading sports drink. The scattergram in Display 4.56 contains the history of the monthly sales of Gaitur Aid during the last three years. The horizontal axis represents the time, in months, starting with January 2002 and ending December 2004. The vertical axis represents the number of bottles sold, in thousands. The manager has asked you to write a summary of the information in this scattergram and to forecast sales for the 12 months of the year 2005. This is your assignment.

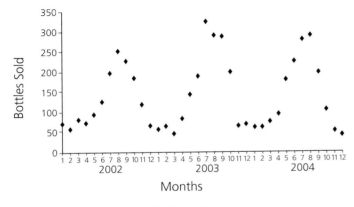

Display 4.56

Appendix A: Using a **TI-84 Plus (TI-83 Plus)** Graphing Calculator

A graphing calculator is a useful tool for doing many different mathematical things. Once you begin to use it, you'll find that it is powerful, fast, and friendly. In fact, your biggest difficulty may be just getting started for the first time! Because this machine can do a lot, it has lots of complicated looking buttons. But you don't have to know about *all* of them before you start to use *any* of them! The sooner you make friends with your electronic assistant, the more it will be able to help you. Let us introduce you to each other by trying a few simple things.

The Cover

The face of the calculator is protected from dirt and scratches by a cover that slides on and off from the top. When you're using the calculator, this cover slips on the back so that you won't lose it. Always put the cover back over the face of the calculator when you finish using it.

On, Off, 2nd , and Clear

To get the calculator's attention, just press ON (at the lower left corner of the calculator). What happens? Do you see a dark block blinking in the upper left corner of the screen? That's the **cursor**, which tells you where you are on the screen. The cursor is always at the spot that will be affected by the next button you push.

Notice that the word [OFF] is printed in blue above the ON button. Notice also that there is one key of the same color. It is the key marked 2nd at the top of the first column of buttons.

When you push 2nd , it makes the next key that you push behave like what is marked above it in blue.

Try it. Push 2nd . What has happened to the cursor? Do you see an up arrow ↑ inside it as it blinks? That's to remind you that 2nd key has been pushed and will affect the next key you choose. Now push ON . What happens? Did the cursor disappear? You should have a blank screen; the calculator should be off.

It's always a good idea to turn your calculator off when you finish using it. If you forget, the calculator will turn itself off after a few minutes to save its batteries. Sometimes when you are using it, you may put it aside and do something else for a little while. If it is off when you pick it up again, don't worry; just press [ON]. The screen will show what was there before it shut down.

Pressing [CLEAR] gives you a blank screen that is ready for new work. But the last thing you did is still stored. Press [2nd] then [ENTER] (bottom right corner) to bring it back.

Basic Arithmetic

Doing arithmetic on a graphing calculator is no harder than on a simpler calculator. In fact, it's easier. This calculator has a screen that lets you keep track of the problem as you enter it. Let's try a few simple exercises. Turn your calculator on.

- Pick two 3-digit numbers and add them. To do this, just key in the first number, press [+] and then key in the second number. Your addition problem will appear on the screen. Press [ENTER] to get the answer.

If you make a mistake when entering a number, you can go back and fix it. The [◄] key lets you move back (left) one space at a time. When you get to your mistake, just key in the correct number over the wrong one. Then move forward (right) to the end of the line by using the [▶] key.

For instance, to add 123 and 456, press 1 2 3 [+] 4 5 6 [ENTER]. The screen will show your question on the first line and the answer at the right side of the second line, as in Display A.1.

```
123+456
            579
```

Display A.1

- Now let's try the other three basic arithmetic operations. To clear the screen, press [CLEAR]. Then try subtracting, multiplying, and dividing your two 3-digit numbers. For instance, if your numbers are 123 and 456, press

1 2 3 [−] 4 5 6 [ENTER]

1 2 3 [×] 4 5 6 [ENTER]

1 2 3 [÷] 4 5 6 [ENTER]

Your screen should look like Display A.2.

Notice that the display uses * for multiplication (so that it is not confused with the letter x) and / for division.

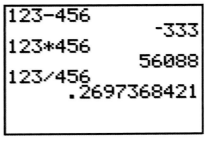

Display A.2

- Here are two button-pushing shortcuts.

If you don't want to redo a problem with just a small change in it, you don't have to reenter the whole thing. [2nd] [ENTER] will bring back the last problem you entered. Just move to the place you want to change, key in the change, and press [ENTER]. For instance, add 54321 and 12345, as in Display A.3.

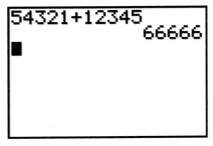

Display A.3

Now, to subtract 12345 from 54321, press [2nd] [ENTER]; the next line will show 54321 + 12345. Move your cursor back to the + sign (using [◄]) and press [−] ; then press [ENTER]. Did you try it? Your screen should look like Display A.4.

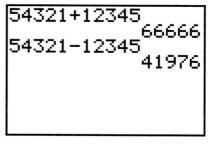

```
54321+12345
              66666
54321-12345
              41976
```

Display A.4

Let's check to see that 41976 is the correct answer by adding 12345 to it and seeing if we get the first number back again. Since you want to do something to the last answer, *you don't have to reenter it*. Press ⊞. Does your calculator show Ans+ and the cursor? It should. If you press an operation key right after doing a calculation, the machine assumes that you want to perform this operation on the last answer. It shows that last answer as Ans. Now key in 12345 and press [ENTER]. You should get back the first number, 54321.

1. Pick two seven-digit numbers and add them. What do you get?

2. Now subtract the second number from the first. Can you do it without rekeying the numbers? What do you get?

3. Now multiply your two seven-digit numbers. What do you get? What does the E mean?

4. Check the last answer by dividing the second of your seven-digit numbers into it. Do it without rekeying the last answer. Do you get your first number back again?

Multiply 98765432 by 123456. Now check the product in two ways.

a

- Divide by pressing ⬚ then entering 123456. Does it check?

- First reenter the product; then divide it by 123456. The product is in scientific notation. To enter it as a regular number, remember that the positive number after the E tells you to move the decimal that many places to the right. Does it check?

1. Divide 97533 by 525 and by 625. One of the answers you get will be exactly right, and the other one will be a very close approximation.

b

  - Which is which?

  - How can you tell?

  - If you hadn't been told that one of the answers is an approximation, how could you know?

2. When an answer is too long to be displayed with ten digits, the calculator shows a ten-digit approximation. Does it do this by just chopping off (truncating) the rest of the digits, or by rounding off? What test would you give your calculator to tell which way it does this?

1. Pick any three-digit number and note it down.

c

2. Repeat its digits in the same order to form a six-digit number (like 123123, for example). Key this number into your calculator.

3. Divide your number by 7.

4. Divide your answer by 11. How do you do this without reentering the answer?

5. Divide the last number by 13. What do you notice about the result? Do you think that it is just a coincidence?

6. Pick another three-digit number and repeat steps 2–5.

7. Try to beat the system; see if you can pick a three-digit number that doesn't work this way. What might you try? Why?

8. Can you actually prove that the pattern you see *works every time*? How might you try to do this?

The Two Minus Signs

The calculator has two minus signs. The one on the gray key looks like ⊟ and the one on the white key looks like ⊡ . The blue one, on the right, is for subtraction. It is grouped with the keys for the other arithmetic operations. To subtract 3764 from 8902, for example, you would key in

$$8\ 9\ 0\ 2\ \boxminus\ 3\ 7\ 6\ 4\ \boxed{\text{ENTER}}$$

Go ahead; do it. Do you get 5138?

The white minus key, to the left of the $\boxed{\text{ENTER}}$ key at the bottom, is for making a number negative. It is grouped with the digit keys and the decimal point. To add the numbers -273, 5280, and -2116, for example, you would key in

$$\boxed{(-)}\ 2\ 7\ 3\ \boxed{+}\ 5\ 2\ 8\ 0\ \boxed{+}\ \boxed{(-)}\ 2\ 1\ 1\ 6\ \boxed{\text{ENTER}}$$

Try it. Notice that the display shows these negative signs without the parentheses, but they are smaller and raised a little. To see the difference between this negative sign and the subtraction sign, try subtracting the negative number -567 from 1234. Here are the keystrokes.

$$1234\ \boxminus\ \boxed{(-)}\ 567$$

The display should look like this:

$$1234 - {}^{-}567$$

Raising to a Power

To raise a number to a power, press ⌃ (just above the divsion sign) just before entering the exponent. Thus, to compute 738^5, press

$$7\ 3\ 8\ \boxed{\wedge}\ 5\ \boxed{\text{ENTER}}$$

The screen should look like Display A.5.

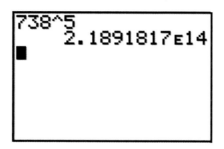

Display A.5

The Menu Keys

Many keys bring a menu to the screen. A menu is a list of functions—things that the calculator is ready to do for you. For instance, press each of the keys across the row that starts with [MATH]. Don't worry about what all those lists say; just pick one out and look at it as you read the rest of this paragraph. Notice that it is actually a double menu. There are two cursors on it, shown as dark blocks. The one in the top left corner can be moved along the top line by using the ◄ and ► keys. Each time you move it to a new place on the top line, the menu below changes. The items in each lower menu are reached by using the other cursor, which can be moved up and down along the left side of the screen by using the ▲ and ▼ keys.

Once you have put the cursor on the choice you want, you actually make the choice by pressing [ENTER]. This makes the calculator go back to its "home" screen and display your choice. To make the calculator do what you have chosen, press [ENTER] again.

1. **How many separate calculator functions can be reached through the menus of the [MATH] key?**

2. **How many separate calculator functions can be reached through the menus of the [MATRIX] key?**

Entering Data in a List

The data handling tools are found through the statistics menu.

- Turn your calculator on and press [STAT]. You'll see a menu that looks like Display A.6.

TI-84 Plus/TI-83 Plus

Display A.6

• To enter data, make sure that the top cursor is on EDIT and the left cursor is on 1: . Then press [ENTER]. Your screen display should look like Display A.7, with the cursor right under L1.

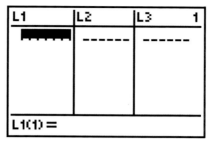

Display A.7

Note that if the display shows numbers in the L1 column, you'll have to clear the data memory. There are two ways to do this.

Get out of this display (by pressing [2nd] [QUIT]) and go back to the [STAT] menu. Press [4]. When ClrList appears, press [2nd] [1] then [ENTER]; Done will appear. Now go back to the [STAT] screen and choose 1:Edit .

or

Without leaving this display, use the [▲] and [◄] keys to move your cursor to the top of the column and highlight L1. Press [CLEAR] and then the [▼] key. List L1 should be cleared.

If you are missing a list in L1 to L6, go back to the STAT menu and press [5] and [ENTER]. This will reset the calculator to List 1 through 6.

• Now it's time to enter the data. Let's use a set of test scores {90,85,95,87,86,92,88,75,81,92} The calculator stores in its memory each data number you enter, along with an L1 label for that entry. The first number is called L1(1), and the second is called L1(2), and so on. We'll ignore the L2 and L3 labels for now. Key in the first data number, then press [ENTER]. Notice that L1(2) now appears at the bottom of the screen. Key in the second data number and press [ENTER]; and so on, until you have put in all the data. If you make a mistake, just use the arrow keys to move the cursor to your error, type over it correctly, then move back to where you were.

At this point, the calculator has all your data stored in a way that is easy to use, and the data will stay stored even after the calculator is turned off.

Summaries of 1-Variable Data

It is easy to get summary information about data that is stored in a single list.

- Bring up the STAT menu.
- Move the top cursor to CALC. The side cursor should be on 1:1-Var Stats. Press [ENTER].
- 1:1-Var Stats will appear on your screen. Enter the list you want the calculator to summarize. For instance, if you want a summary of the data in list L1, press [2nd] [1]; then press [ENTER].

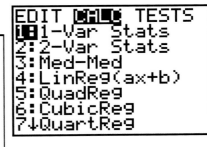

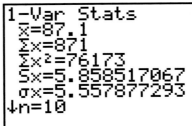

That's all there is to it! A screenful of information will appear. Sections 1.3–1.7 in Chapter 1 of Year 1 explain how to interpret that information.

Putting Data in Size Order

The TI-84 Plus (TI-83 Plus) have built-in programs that will put your data in size order automatically. Let's order the test scores that you just entered.

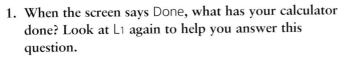

1. Go to the STAT menu and choose 2:SortA(then press [ENTER].

2. Tell the calculator to sort the L1 list by pressing [2nd] [1] then [ENTER]. Your calculator screen should now say DONE.

3. To see what it has done, reopen List 1 (using [STAT] 1:Edit). Your data should now be listed in ascending order — that is, from smallest to largest as you read down the list. The A in SortA(stands for ascending order

Now go back to the STAT menu and choose 3:SortD(then press [ENTER]. Tell the calculator to sort the L1 list again. (Press [2nd] [1] [ENTER].)

1. When the screen says Done, what has your calculator done? Look at L1 again to help you answer this question.

2. What does the "D" in SortD(stand for?

Once the data are in size order, it is easy to find the median. For example, if you have 21 data items in all, the median is just the 11th one in the sorted list. Scroll through the data (using the ▲ key) until you find L1(11). Its value is the median. If you have 20 data items, the median is halfway between the 10th and 11th items in the sorted list. Scroll through the data until you find L1(10) and L1(11). Then calculate the number halfway between them.

Finding the mode is just as easy. Count repeated items in this list. The one that is repeated the most times is the mode.

The Graph Window

This kind of calculator is called a graphing calculator because it can *draw graphs*. The screen on a graphing calculator can show line drawings of mathematical relationships. It does this with two kinds of coordinate systems—*rectangular coordinates* or *polar coordinates*. In this part we shall use only rectangular coordinates; polar coordinates will appear much later. If you are not familiar with the idea of a rectangular coordinate system, you should review the first section of Chapter 3 in Year 1 now.

Your calculator leaves the factory with standard coordinate axes built in. To see what they look like, turn on your calculator and press GRAPH (in the upper right corner). You should see a horizontal and a vertical axis crossing the middle of the screen. The horizontal axis is called the *x*-axis, and the vertical axis is called the *y*-axis. If your screen doesn't show this, press ZOOM and choose 6:ZStandard . Examine this display carefully; then answer the following questions.

1. Assuming that the dots along each axis mark the integer points, what is the largest possible value on the *x*-axis? On the *y*-axis?

2. What is the smallest possible value on the *x*-axis? On the *y*-axis?

3. Does it look as if the same unit of measure is being used on both axes?

4. Why do you suppose the spacing between the units is not exactly the same everywhere on an axis? Do you think that this might cause a problem?

The standard coordinate axis setting can be changed in several ways. This is done using the menu that appears when you press WINDOW. Try that now. You should get Display A.8.

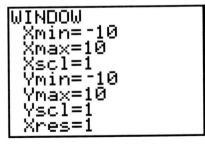

TI-84 Plus/TI-83 Plus

Display A.8

Xmin and Xmax are the smallest and largest values on the x-axis (the horizontal axis); Ymin and Ymax are the smallest and largest values on the y-axis (the vertical axis).

Xscl and Yscl are the scales for marking off points on the axes. The setting 1 means that each single integer value on the axis is marked. To see how the scale value works, change Xscl to 2. Move the cursor down, using ▼, then just key in 2 in place of 1. Now press GRAPH. What change do you notice? Now go back to the WINDOW menu (press WINDOW) and change Yscl to 5. Return to the graph (press GRAPH). What has changed?

The Xres = 1 line indicates the resolution of the graph. It can be set to an integer from 1 to 8. At Xres = 1, it evaluates a function at each of the 94 pixels on the x-axis. At Xres = 8, it evaluates the function at every eighth pixel.

Change the WINDOW **settings so that they look like Display A.9. Then look at the graph and answer these questions.**

1. **Where on the screen is the origin of the coordinate system?**

2. **Does it look as if the same unit of measure is being used on both axes?**

3. **Does it look as if the spacing between the units is the same everywhere on an axis?**

4. **What happens when you press ▲ then ▼?**

```
WINDOW
 Xmin=0
 Xmax=9.4
 Xscl=1
 Ymin=0
 Ymax=6.2
 Yscl=1
 Xres=1
```

TI-84 Plus/TI-83 Plus

Display A.9

If you have worked through the previous questions, you found that pressing ▲, ▼ puts a cross exactly in the middle of your screen and two numbers at the bottom. The cross is the cursor for the graphing screen, and the numbers are the coordinates of the point at its center. In this case, the cursor is at (4.7, 3.1). It can be moved to any point on the graph by using the four arrow keys ▲, ▼, ◄, ► at the upper right of the keypad.

Move the cursor to the point (4, 3). How far does the cursor move each time you press ◄ or ►? How far does it move each time you press ▲ or ▼? Now move the cursor directly down to the bottom of the screen. What are the coordinates of the lowest point you can reach?

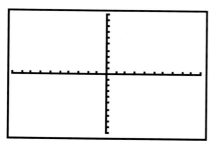

These new WINDOW settings are better than the standard one in some ways, and worse in others. Let's look again at the Standard coordinate system and compare it with the one we just saw. To get back to the standard settings, press [ZOOM], then press [6] to choose ZStandard. The Standard coordinate axes should appear immediately.

These questions refer to the Standard coordinate axes.

1. Where is the cursor to begin with? How do you find it if you can't remember?

2. Try to move the cursor to the point (4,3). How close can you get to it?

3. How far does the cursor move each time you press ◄ or ►?

4. How far does it move each time you press ▲ or ▼?

5. Move the cursor directly down to the bottom of the screen. What are the coordinates of the lowest point you can reach?

6. In what ways is this coordinate system better than the one we set up for the previous set of questions? In what ways is it worse?

7. How might we fix the bad features of this system without losing the good ones?

Another useful WINDOW setting is 8:ZInteger in the ZOOM menu. When you press 8, coordinate axes appear, but they are still the standard ones. Press ENTER to get the Integer settings.

These questions refer to the Integer coordinate axes.

1. Try to move the cursor to the point (4, 3). How close can you get to it?

2. How far does the cursor move each time you press $◄$ or $►$?

3. How far does it move each time you press $▲$ or $▼$?

4. Why is this setting named Integer?

To plot a point (mark its location) on the graphing screen, go to the point-drawing part of the DRAW menu, like this.

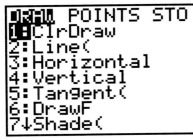

Draw Menu   Points Menu

Press 2nd PRGM and move the top cursor to POINTS.

Choose 1 to make the ENTER key mark cursor locations. If you want to mark some points and erase others, choose 3. This lets the ENTER key change the state of any point the cursor is on; it will mark one that isn't already marked, and will unmark one that is. *Hint:* If you have plotted too many points and you want to start over, you can go to Draw menu and enter 1 for ClearDraw.. This will wipe out everything you have plotted and return to the Standard coordinate settings. If you were using different coordinate settings, you will have to redo them in the WINDOW menu. If you want to erase some points, see Drawing Points in the graph section of TI Guidebook.

Problem Set: Appendix A

1. What WINDOW settings do you need in order to put the origin at the upper right corner of your screen? What can you say about the coordinates of the points that can be plotted on this screen?

2. What WINDOW settings do you need in order to put the origin at the upper left corner of your screen? What can you say about the coordinates of the points that can be plotted on this screen?

3. Choose the Integer setting for the coordinate axes and plot the points (30, 14), (-5, 20), (-26, -11), and (6, -30). Then write the coordinates of two points that lie within the area of the graph window but cannot be plotted exactly with this setting.

4. Find WINDOW settings to form a coordinate system such that the points (120, 80) and (-60, -40) are within the window frame.

   (a) How far does the cursor move each time you press ◄ or ►?

   (b) How far does it move each time you press ▲ or ▼?

   (c) Can you put the cursor exactly on (120, 80)? If not, how close can you come? Plot this point as closely as you can.

   (d) Can you put the cursor exactly on (-60, -40)? If not, how close can you come? Plot this point as closely as you can.

   (e) Can you put the cursor exactly on (0, 0)? If not, how close can you come?

5. Find WINDOW settings to form a coordinate system such that the cursor can be put exactly on the points (20, 24.5) and (-17.3, -14).

   (a) What is the initial position of the cursor?

   (b) How far does it move each time you press ◄ or ►?

   (c) How far does it move each time you press ▲ or ▼?

   (d) Can you put the cursor exactly on (0, 0)? If not, how close can you come?

Drawing Histograms

Drawing a histogram is very easy. All you have to do is choose a few numbers to tell the calculator how wide and how tall to make the bars, as follows. Turn your calculator on and press WINDOW. The screen should look like Display A.10, maybe with different numbers.

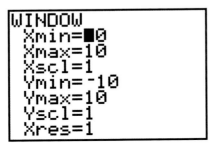

TI-84 Plus/TI-83 Plus

Display A.10

The numbers in this WINDOW list tell the calculator how to set the horizontal (X) and vertical (Y) scales.

- Xmin, an abbreviation of *X minimum*, is the smallest data value the picture will show. You should set it at some convenient value less than or equal to the smallest value in your data set.

- Xmax, an abbreviation of *X maximum*, is the largest data value the picture will show. Set it at some convenient value greater than or equal to the largest value in your data set.

- Xscl, an abbreviation of *X scale*, says how to group the data. It is the size of the base interval at the bottom of each bar of the histogram. For instance, Xscl = 10 will group the data by 10s, starting from the value of Xmin that you chose.

- Ymin is the smallest frequency of any data group. It is never less than 0, which usually is a good choice for it.

- Ymax represents the length of the longest bar. Choose a convenient number that is not less than the largest frequency of any data group, but not much larger.

- Yscl determines the size of the steps to be marked on the vertical (frequency) scale. For small data sets, set it to 1. If your setting for Ymax is much larger than 10, you might want to set Yscl larger than 1. A little experimenting will show you how to choose a helpful setting.

- Leave Xres=1.

Now your calculator is ready to draw a histogram.

- Press [STAT PLOT] (actually, [2nd] [Y=]), choose 1 and press [ENTER].

- Choose these settings from each row by moving the cursor to them with the arrow keys and pressing [ENTER] each time.

  – Highlight On.

  – Highlight the histogram picture.

  – Set Xlist to the list containing your data (L1, L2, etc.).

  – Set Freq:1 .

Now press [GRAPH] — and there it is!

Drawing Boxplots

The TI-84 Plus (TI-83 Plus) calculators can draw boxplots. All they need are the data and a few sizing instructions. Here's how to do it.

- Turn the calculator on, press [STAT] and choose 1:Edit... from the EDIT menu. Check which list contains the data you want to use. Let's assume it's in L1.

- Press [WINDOW] and set the horizontal (X) and vertical (Y) scales. If you have forgotten how to set your WINDOW, refer to "The Graph Window" section. Choose convenient numbers for the X range—Xmin less than your smallest data value and Xmax greater than your largest data value, but not too small or too large. You don't want the picture to get squeezed into something you can't see well! Also set Xscl to some convenient size.

- The Y settings don't matter as much. Ymin = -2 and Ymax = 3 work a little better.

- Press [STAT PLOT] (actually, [2nd] [Y=]), choose 1: and press [ENTER]. Select these settings from each row by moving the cursor to them with the arrow keys and pressing [ENTER] each time.

  On; the boxplot picture; L1 from the Xlist; 1 from Freq

- Now press [GRAPH] —and there it is!

- To read the five-number summary, press [TRACE] and use the [◄] and [►] to display the five numbers one at a time.

Graphing and Tracing Lines

If you want the calculator to graph a line or a curve, you must first be able to describe the line or curve by an algebraic equation. Once you have the equation for what you want to draw, you must put it in the form

$$y = [\text{something}]$$

For a straight line, that's not a problem; we often put the equation in this form, anyway. For some other kinds of curves, putting them in this form can be a little messy. In this section we shall deal only with straight lines.

All graphing begins with the $\boxed{Y=}$ key. When you press this key for the first time, you get the screen in Display A.11.

TI-84 Plus/TI-83 Plus

Display A.11

These lines allow you to put in as many as ten different algebraic equations for things you want drawn. The subscript number gives you a way to keep track of which equation goes with which picture on the graph. To see how the process works, we'll make the first example simple—two straight lines through the origin.

> Key in -.5X on the $Y_1=$ line, *using the* $\boxed{X,T,\Theta,n}$ *key to make the X*; then press $\boxed{\text{ENTER}}$.

It is important to use $\boxed{X,T,\Theta,n}$ for X because that's how the calculator knows that you are referring to the horizontal axis.

Key in -.25X on the $Y_2=$ line and $\boxed{\text{ENTER}}$ it.

Be sure to use the $\boxed{(-)}$ key for the negative sign. If you don't, you'll get an error message when you ask for the graph. If you want to wipe out one of these equations and redo it, just move the cursor back to the equation and press $\boxed{\text{CLEAR}}$.

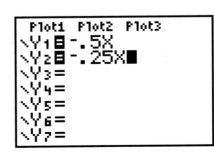

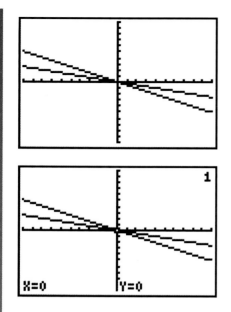

Now your work is done. Press GRAPH and just watch as the calculator draws the lines. If you forget which line goes with which equation, or if you want to see the coordinates of the points along your lines, press TRACE and then move the cursor with the ◄ and ► keys. When you do this, the coordinates of the cursor's position appear at the bottom of the screen. The equation appears in the upper left hand corner. In this example, when you press TRACE you will be on $Y_1 = -.5X$, the first of the two lines we entered. Try it. To switch from one line to another, use the ▲ and ▼ keys. Notice that, in this case, either of these keys gets you to the other line. That's because we are only graphing two equations. If we were graphing more than two, these keys would move up and down the *list of equations*, regardless of where the graphs appeared on the screen.

There is a way to remove the graph of an equation from the screen without erasing the equation from your list. For example, let us remove the line Y1 = -.5X from the picture. Go back to the Y= list. Notice that the = sign of each equation appears in a dark block. This shows that the graph of this equation is turned on. To turn it off, move the cursor to the = sign and press ENTER. The dark block will disappear. To turn it back on, put the cursor back on = and press ENTER again.

Approximating Data by a Line

This section refers to a situation that commonly arises in the analysis of two variable data. Such data can be represented as points on a coordinate plane, and it is often useful to know if the pattern of points can be approximated by a straight line. A common way of doing this is called *least-squares approximation*. An explanation of this process and its use appears in Chapter 4 of Year 1. This calculator section provides a simple example of how to get the TI-84 Plus (TI-83 Plus) to give you a least-squares approximation of a set of data.

Let's look at a very small, simple data set. The process is exactly the same for bigger, more complicated data sets. Here are four points of two variable data.

$$(1, 2) \quad (2, 3) \quad (3, 5) \quad (4, 6)$$

If you plot these points on a coordinate plane, you will see that they don't all lie on the same line. Don't just take our word for it; make a sketch! The calculator uses the least-squares method to find automatically the line of "best fit."

Section 4.3 (in Chapter 4) describes how this method works and what best fit means. These are the instructions for getting the calculator to do all the tedious work for you.

First of all, you need to have the data entered in two *separate data storage lists.* You get to these lists by pressing [STAT] and choosing 1:EDIT... from the EDIT menu. When you press [ENTER], you should get Display A.12.

Display A.12

If the columns already contain data that you don't want, you can clear them out in either of two ways.

- Press [STAT] and choose 4:ClrList from the menu that appears. When the message ClrList appears, enter the name of the list you want to clear. (Press [2nd] [1] for L1, [2nd] [2] for L2, etc.) Then press [ENTER]; the screen will say Done. Now press [STAT] to return to the process of entering data.

- Go to the first element in your list L1(1) and press [DEL]. Continue to do this until the list is gone. This is a reasonable method as long as the list is short.

Enter the first coordinate of each data point into list L1; put its second coordinate in list L2. The four data points of our example should appear as shown in Display A.13.

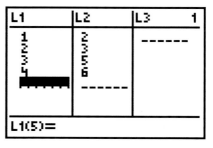

Display A.13

Now we are almost done. Press [STAT] and go to the CALC menu. Choose LinReg(ax + b). When you press [ENTER], the screen will display an algebraic description of the line of best fit. For our example, it looks like Display A.14.

- The second line, y = ax + b, just tells you that the information is for slope-intercept form. Notice that the TI-84 Plus (TI-83 Plus) use *a*, not *m*, for the slope here.

- The third line says that the slope is 1.4.

- The fourth line says that the *y*-intercept is .5.

Note that the last two lines may show the **correlation coefficient *r*, and the coefficient of determination *r^2*. The correlation coefficient is not discussed in your textbook. A detailed explanation of how it works will have to wait until you study statistics in more depth. But, in case you are curious about it, here is a little more information. The correlation coefficient is always a number between -1 and 1, inclusive. 1 and -1 stand for a perfect fit, with all points exactly on the line. (1 is for lines with positive slope; -1 is for lines with negative slope.) The closer *r* is to 0, the worse the fit.

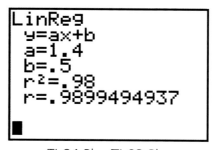

TI-84 Plus/TI-83 Plus

Display A.14

Putting together this information about our example, we see that the least-squares line is described by the equation

$$y = 1.4x + .5$$

Graph the line $y = 1.4x + .5$. Are any of the four data points on it? How can you be sure?

Using Formulas to Make Lists

Sometimes it is useful to make a new list of data from an old one by doing the same thing to each data value. For instance, you might want to add a fixed number to each value, square each value, or find the distance of each value from some particular number. Instead of computing the new list one entry at a time, you can do it all at once if you can express your process as a formula.

Here's how the process works.

- Go to the STAT menu. Enter a list of data in L1, and then clear L2 and L3.

- To add 5 to each entry in L1, move the cursor over to the second column, then up to the heading, L2. The bottom line of your display should read L2= (without any number in parentheses).

- The trick here is to let the symbol L1 stand for each element of the list L1. That is, we make L1 *a variable*. Key in L1 + 5 ; the bottom of your screen should read L2 = L1 + 5 .

- Now press [ENTER] and watch the entire column for L2 fill out automatically!

- To list in L2, the square of each entry in L1, put the cursor on L2 (at the top of the column). Then enter L1^2 (or L1 * L1).

- Now let us list in L3 the midpoint between the L1 entry and the L2 entry. Put the cursor back on L3 at the top of the column and press [CLEAR]. This removes the old formula. Now key in (L1 + L2)/2 and press [ENTER].

    1. **List at least ten data values in L1.**

    2. **Write a formula to list in L2 the distance between 17 and each entry in L1. Remember: Distances are never negative numbers. Then use it.**

3. Write a formula to list in L3 the square of the difference (which may be negative) between each entry in L_1 and 17. Then use it.

4. Write a formula to list in L4 the square root of each entry in L3. Then use it.

5. How are columns L2 and L4 related? Explain.

Drawing Circles

To draw circles directly on a graph, use 9:Circle(in the DRAW menu. (The DRAW menu appears when you press [2nd] [DRAW].) 2:Line(can be used to draw segments, which lets you add radii, diameters, and other segments to your drawings of circles.

Before beginning, make sure that all the functions on your Y= screen are turned off. If they are not, their graphs will appear when you draw circles and segments. Also make sure that all STAT PLOTS are turned off.

Follow these instructions to draw a circle directly on a graph.

1. From the ZOOM menu, choose ZStandard (to clear any unusual WINDOW settings). Then choose ZSquare or ZInteger, which displays the graph window.

2. From the DRAW menu, choose 9:Circle(.

3. Choose a point for the center by moving the cursor to this point and pressing [ENTER].

4. Choose the radius for your circle by moving the cursor this many units away from the center and pressing [ENTER]

You can continue to draw circles by repeating the last two steps. To clear the screen before drawing a new circle, use :ClrDraw in the DRAW menu. If you want to stop drawing circles, press [CLEAR].

Follow the steps above to draw each of these items.

1. a circle with center (0, 10) and radius 5

2. a circle with center (12, −7) and radius 15

3. four circles with center (0, 0)

You can also draw a circle from the Home Screen (the calculator's primary display WINDOW) by following these instructions. You can use this same method to draw circles from a program.

1. From the Home Screen, choose Circle(from the DRAW menu.

2. Input the coordinates of the center, followed by the radius; then press ENTER. For example, if you enter (0, 10, 5), the calculator will draw a circle with center at (0, 10) and radius 5, using whatever ZOOM WINDOW setting is current.

3. To return to the Home Screen, press CLEAR.

1. Draw a circle with center (3, 2) and radius 7 directly from the Home Screen. If your graph does not look like a circle, how can you adjust the graph WINDOW so that it does?

2. Draw four concentric circles around (0, 0) directly from the Home Screen. Earlier you were asked to draw this figure directly on a graph. Which method is easier for you? Why?

Appendix B: Using a Spreadsheet

Appendix B:
Using a Spreadsheet

Computers give us many different tools for doing and using mathematics. One of these tools is called a **spreadsheet**. These days, a spreadsheet is an easy-to-use and very powerful computer program, but the idea of a spreadsheet is really much simpler and older than computers. Originally, a spreadsheet was just an oversized piece of paper, with lines and columns that made it easier for accountants and bookkeepers to keep their work in order.

You can make a spreadsheet on a lined piece of paper.

- Make a narrow border across the top and down the left side of the sheet.

- Divide the rest of the paper into columns from top to bottom. Six columns of about equal width will do for now.

- In the left margin, number the lines, beginning with 1, to the bottom of the page.

- Across the top margin, name each column with a letter from A to F in alphabetical order.

Your paper should look something like Display B.1.

	A	B	C	D	E	F
1						
2						
3						
4						
5						
6						
⋮						

Display B.1

The Cell Names

Each box in this grid has its own address — the letter of its column followed by the number of its row. For instance, C4 refers to the box, third column (column C), on the fourth line (row). The electronic spreadsheets that computers handle look just like this, and each position in them is addressed in just the same way. Electronic spreadsheet manuals often call the boxes **cells.** We'll do the same thing, so that you become used to the term.

> Here are a couple of questions to get you comfortable with the way cells are addressed.
> - Make a copy of Display B.1 and shade in these cells: A2, B3, C4, D5, E6, A6, B5, D3, E2. What shape do you get?
>
> - If you wanted Display B.1 to be shaded in a checkerboard pattern, with alternating cells filled in, which cells would you shade? Write out all their addresses. There's more than one way to do this.

The advantage of electronic spreadsheets over handmade ones is that the electronic ones do the computations for you, *IF* you ask them properly. If you know how to speak the language of your spreadsheet program, you can get it to do all the hard work very quickly. The main idea to remember is

> A spreadsheet is powerful because it can find and work with numbers that appear anywhere on it by using the cell names.

Therefore,

> When working with a spreadsheet, always try to build what you want, step-by-step, from the first data you enter. The fewer numbers you have to enter, the easier it is for the spreadsheet to do your work.

The rest of this appendix shows you how to get an electronic spreadsheet to work for you. For practice, each new process will be introduced by using it to deal with this problem.

> You are sent to the local supermarket to buy at least 2 pounds of potato chips for a club picnic. The club treasurer tells you to spend as little money as possible.

Now, there are many different brands of potato chips, and each brand comes in several different size bags. How can you compare prices in a useful way? Well, the bag sizes are measured in ounces. If you divide the price of the bag by the number of ounces, you'll get the price per ounce (this approach is called *unit* pricing). We'll set up a spreadsheet to tell you the price per ounce of every kind of potato chip bag your market sells.

> *Don't just read the rest of this appendix*: **DO IT! Work along with the instructions using your own spreadsheet.**

Entering Numbers and Text

There are three different kinds of things you can put in a cell — numbers, text, and formulas. Most spreadsheets distinguish between numbers and text automatically.

1. If you enter numerical symbols only, the entry is treated as a number.

2. If you begin an entry with letters or other symbols not related to numbers (even if numbers are entered along with them), the entry is treated as text.

Note that if you want a number (such as a date or a year) or a number related symbol (such as $) to be treated as a text entry, you have to tell the machine somehow. Check your user's manual for the way your spreadsheet program does it.

Display B.2 lists the prices of different brands and sizes of potato chips, including the special sale prices for the day. These are actual data from a supermarket. To enter these data in their most useful form, you should use *three* columns—one for the brand, one for the weights (in ounces), and one for the prices. Put the information of Display B.2 into columns A, B, and C now.

Brand of Chip	No. of Ounces	$Cost of Bag
Cape Cod	11 oz.	2.49
Eagle Thins	9.5 oz.	1.99
Humpty Dumpty	6 oz.	1.19
Humpty Dumpty	10 oz.	1.68
Lay's	6 oz.	0.95
Lay's	14 oz.	2.79
O'Boisies	14.5 oz.	2.79
Ruffles	6 oz.	1.39
Ruffles	14 oz.	2.79
Tom's	6 oz.	1.39
Tom's	11 oz.	1.69
Wise	6 oz.	1.39
Wise	10 oz.	1.48

Display B.2

The standard column width of your spreadsheet probably is not big enough to handle some of the brand names. Find the Column Width command and adjust the width of column A to 15 spaces. While you're at it, you might as well adjust the width of column B (the ounces) and column C (the price) each seven spaces wide. This will make the display look a little neater.

Entering Formulas

If you want the spreadsheet to calculate an entry from other data, you have to give it a formula to use. You also have to begin with a special symbol to let it know that a formula is about to be entered. The special symbol depends on the type of spreadsheet you have. Excel uses the symbol = ; Lotus 1-2-3 uses the symbol + ; your software might use something else.

Calculate the price per ounce of Cape Cod chips by entering the formula C1/B1 into cell D1. As soon as you enter it, the number 0.226363 should appear. This is correct, but more accurate than we need. Three decimal places should be enough. Find the spreadsheet command that fixes the number of decimal places and use it to set the column D display to 3 places.

Copying Formulas

To get the price per ounce of Eagle Thins, all you have to do is copy the formula from cell D1 to cell D2. Do that. Check the spreadsheet manual to see how to copy from one cell to another. As soon as you do it, the number 0.209 will appear. Now look at the formula itself. Notice that it says C2/B2; that is, when you copied the formula one cell below where it started, the spreadsheet automatically changed the cell addresses inside it by that amount. This automatic adjustment process is one of the most powerful features of the spreadsheet. Next we'll use it to get the price per ounce of *all* the other kinds of chips at once!

Repeated Copying

You can copy a cell entry over and over again, all at once, along as much of a row or column as you mark out. If the entry is a formula, the spreadsheet will automatically adjust the cell addresses in it at each step. In some spreadsheet programs (such as Excel), this is done by the Fill command. In others (such as Lotus 1-2-3), it is done as part of the Copy command, by highlighting the entire region of cells into which you want the formula copied.

Find out how this works for your spreadsheet. Then copy what's in D2 into cells D3 through D13 and watch all the per ounce prices appear immediately. At this point, your spreadsheet should look something like Display B.3.

	A	B	C	D
1	Cape Cod	11	2.49	0.226
2	Eagle Thins	9.5	1.99	0.209
3	Humpty Dumpty	6	1.19	0.198
4	Humpty Dumpty	10	1.68	0.168
5	Lay's	6	0.95	0.158
6	Lay's	14	2.79	0.199
7	O'Boisies	14.5	2.79	0.192
8	Ruffles	6	1.39	0.232
9	Ruffles	14	2.79	0.199
10	Tom's	6	1.39	0.232
11	Tom's	11	1.69	0.154
12	Wise	6	1.39	0.232
13	Wise	10	1.48	0.148

Display B.3

What *formula* is being used in cell D3? In D7? In D13?

Inserting Rows and Columns

Now let's put in column headings so that the spreadsheet is easier to understand. Move the cursor to the beginning of row 1 and use the Insert Row command of your spreadsheet to put in two rows at the very top. Cape Cod should now be in cell A3. We'll use the first row for headings and leave the second row blank. Enter Brand in A1, ounce in B1, price in C1 and enter $/oz. in D1. Change the width of column D to 7 spaces.

Because we've moved everything down, the row numbers no longer correspond to the number of brands listed. Make space to renumber the rows that list the brands, like this: Move the cursor to the top of the first column and use the Insert Column command to put two new columns at the far left. Cape Cod should now be in cell C3.

Numbering Rows

Now let's try a little experiment. We'll number the brands in two different ways. Make the two new columns, A and B, only 4 spaces wide. Now put the numbers 1 through 13 down these two columns, starting at the third row, in these two ways.

- In column A, enter each number by hand—the number 1 in A3, the number 2 in A4, and so on, down to the number 13 in A15.

- In column B, enter the formula B2+1 in cell B3. The number 1 will appear because the spreadsheet treats the empty cell B2 as if it had 0 in it. Now copy this formula into all the cells from B3 through B15.

Do columns A and B match? They should. If they don't, ask your teacher to help you find what went wrong. At this point, your display should look like Display B.4.

	A	B	C	D	E	F
1			Brand	oz.	$/bag	$/oz.
2						
3	1	1	Cape Cod	11	2.49	0.226
4	2	2	Eagle Thins	9.5	1.99	0.209
5	3	3	Humpty Dumpty	6	1.19	0.198
6	4	4	Humpty Dumpty	10	1.68	0.168
7	5	5	Lay's	6	0.95	0.158
8	6	6	Lay's	14	2.79	0.199
9	7	7	O'Boisies	14.5	2.79	0.192
10	8	8	Ruffles	6	1.39	0.232
11	9	9	Ruffles	14	2.79	0.199
12	10	10	Tom's	6	1.39	0.232
13	11	11	Tom's	11	1.69	0.154
14	12	12	Wise	6	1.39	0.232
15	13	13	Wise	10	1.48	0.148

Display B.4

Ordering Data

Another handy feature of an electronic spreadsheet is that it can put in order data that is listed in a column. It can put numbers in size order, either increasing or decreasing. Most spreadsheets can also put text entries in alphabetical order. To do this, you need to find the Sort command and tell it what list of data you want to rearrange. In Excel, Sort is in the Data menu; in Lotus 1–2–3, it's in the Select menu. The computer prompts you for a little more information, such as whether you want ascending or descending order, then does the sorting.

Note that some spreadsheets move entire rows when they sort; others can be told just to rearrange the data in a single column. Check your user's manual to see how your spreadsheet works. In this example, we assume that the spreadsheet moves entire rows when it sorts.

Let's rearrange the potato chip list according to the price per ounce, from most expensive to least expensive. Follow your spreadsheet's instructions to sort the per ounce prices in column F in ascending order. Which kind of potato chip is the best buy? Which is the worst buy?

> Look at columns A and B.
>
> 1. Do they still match? What has happened? Explain.
>
> 2. What would have happened if you had entered the number 1 in B3, then entered the formula =B3+1 in B4? Explain.

Now that we have all this information, how do we find out how much it will cost the club for the 2 pounds of potato chips? Here's one plan.
- Compute the number of ounces in 2 pounds.
- Multiply the cost of 1 ounce by the total number of ounces needed.

Warning: There's something wrong with this approach; what is it?

> We'll do this on the spreadsheet because it provides an example of a different way to use cell addresses. To find the total number of ounces, we just multiply the number of pounds (2) by 16. Make these entries on the spreadsheet.

- In C17, enter number of lbs.; in D17, enter 2.

- In C18, enter number of oz.; in D18, put the formula that multiplies the entry in D17 by 16. (What is that value?)

Constant Cell Addresses

To find out how much 2 pounds of each kind of potato chip will cost, first set column G to display in currency format. Then move the cursor to cell G3. This should be the first blank cell at the end of row 3. We want this cell to show the number of ounces to be bought (in D18) multiplied by the price per ounce (in F3). Let's try it.

- Enter the formula D18*F3 in G3. The result should be $4.74.
  Is your first kind the Wise 10 oz. bag?

- So far, so good. Now copy this formula to the next line, in G4. What do you get? $0.00? How come?

- Look at the formula as it appears in G4. Does it say D19*F4? What happened?

Remember that when you shift a formula from one location to another, the spreadsheet automatically shifts every cell address in exactly the same way. We copied this formula to a location one row down from where it was, so the spreadsheet added the number 1 to the row number of each cell address in the formula. Now, we want that to happen to one of these addresses, but not to the other. That is, the cost of the kind of potato chip in row 4 should use the price per ounce in F4, but it should still use the total number of ounces from D18.

 To prevent the spreadsheet from automatically adjusting a cell address when a formula is moved, enter the cell address with a $ in front of its column letter and a $ in front of its row number.

This means that you should go back to cell G3 and enter the multiplication formula $D$18*F3. Now copy this to G4. Do you get $4.92? Good. If not, what went wrong? Ask your teacher if you need help figuring it out. Now copy this formula into cells G5 through G15. Column G now should show the cost for 2 pounds of each kind of potato chip in your list.

Go to G1, make this column 7 spaces wide and enter the word cost as the column heading.

1. According to column G, which kind of potato chip is the best buy?

2. Why is that *not* necessarily the best buy for your club?

3. What's wrong with letting this answer tell you what kind to buy? *Hint*: How many *bags* would you have to buy?

The INT Function

As the hint in the box above suggests, using the information in column G to guide your choice may not be a good idea because the supermarket sells potato chips by the bag. In order to know how much it will cost to get at least 2 pounds of chips, you first must know how many bags you'll need.

How do you do that? Easy, right? Just divide 32 oz. (2 lbs.) by the number of ounces in a single bag. If you get a mixed number, add 1 to the whole number part.

For example, if you want at least 32 oz. in 10 oz. bags, divide 32 by 10. You get 3.2 as an answer, but, since you can't buy 0.2 of a bag of chips, you need 4 bags.

There's a spreadsheet function—called INT—that makes this very easy to compute automatically. The INT function gives you the greatest integer less than or equal to the number you put into it. For instance.

$$\text{INT}\left(3\tfrac{1}{3}\right) = 3$$

$$\text{INT}(2.98) = 2$$

$$\text{INT}(5) =$$

Let's use this function to carry out the computation we just did, finding how many 10 oz. bags of Wise potato chips we need in order to have at least 2 pounds. But instead of entering the numbers in separately, we'll get them from other cells on the spreadsheet. Move to cell H3 and enter the formula

$$\text{INT}(\$D\$18/D3) + 1$$

Just to make sure you understand what we're doing, answer these questions before moving on.

1. What does $D$18 stand for?

2. Why are the $ symbols there?

3. What does D3 stand for?

4. What number is $D$18/D3?

5. What number is INT($D$18/D3)?

6. What number is INT($D$18/D3)+1?

7. If you copy this formula to cell H4, how will it read?

Now use the Fill command to copy this formula into cells H4 through H15. For each kind, the number you get says how many bags you need in order to have at least two pounds of chips. Put the heading bags at the top of this column H, and make the column 5 spaces wide.

Now we can finish the problem. To find the cost of at least 2 lbs. of each kind of chip, multiply the number of bags you need by the cost of a single bag. Enter a formula in I3 that does this; then copy it into I3 through I15. Finish your spreadsheet display by renaming column G cost 1 and naming column I cost 2 and changing the width of column I to 6 spaces.

1. If you *must* get at least 2 lbs. of chips and you want to spend as little as possible, which kind do you buy?

2. How many bags do you buy?

3. What does it cost you?

The next questions show the power of spreadsheets for testing out different variations of a situation. Each part is exactly the same as above, except that the total number of pounds of chips is different. Answer each one by changing as little as possible on your spreadsheet.

1. If you *must* get at least 3 lbs. of chips and you want to spend as little as possible, which kind do you buy? How many bags do you buy? What does it cost you?

2. If you *must* get at least 4 lbs. of chips and you want to spend as little as possible, which kind do you buy? How many bags do you buy? What does it cost you?

3. If you *must* get at least 5 lbs. of chips and you want to spend as little as possible, which kind do you buy? How many bags do you buy? What does it cost you?

Problem Set: Appendix B

1. These two questions refer to the potato chip spreadsheet that you just made.

   (a) Add a column J that shows the total number of ounces of potato chips of each kind that you get when you buy enough bags to get at least two pounds of potato chips. What formula will compute these numbers?

   (b) Add a column K that shows the number of *extra* ounces (more than 2 pounds) that you get when you buy enough bags to get at least two pounds of potato chips. What formula will compute these numbers?

2. Make a spreadsheet like the one for the potato chips to deal with this problem.

   (a) Your favorite aunt runs a shelter for homeless cats. As a present for her birthday, you decide to give her 5 pounds of canned cat food. You want to spend as little money as possible. The brands, sizes, and prices for the canned cat food at the supermarket are shown in Display B.5. What brand and size is the best buy, and how many cans of it should you get? What will it cost?

   (b) Your best friend thinks you have a great idea. She decides to buy your aunt 5 pounds of canned cat food, too. If you both chip in and buy a combined present of 10 pounds of canned cat food, what is the best buy of canned cat food for your combined present? Explain your answer.

Brand of Cat Food	No. of Ounces	Cost
Alpo	6 oz.	3 for $1.00
Alpo	13.75 oz.	$ .65
Figaro	5.5 oz.	$ .37
Figaro	12 oz.	$ .66
Friskies	6 oz.	$ .35
Friskies	13 oz.	$ .58
Kal Kan	5.5 oz.	4 for $1.00
Puss 'n Boots	14 oz.	$ .55
9 Lives	5.5 oz.	3 for $ .88
9 Lives	13 oz.	$ .48
Whiskas	5.5 oz.	3 for $1.00
Whiskas	12.3 oz.	$ .55

Display B.5

3. Here's a bonus question.

   (a) Invent a problem about breakfast foods that is like the potato chip and cat food problems.

   (b) Go to your local supermarket and gather the brand, size, and price information that you will need to solve your problem.

   (c) Using the data you gather for part (b), set up a spreadsheet that solves the problem you invented in part (a).

Appendix C: Programming the TI-82 (TI-83)

After you have been using the TI-82 (TI-83) for a while, you may notice that you are repeating certain tasks on your calculator over and over. Often you are repeating the same sequence of keystrokes, which can become very tiresome. Programs give you a way to carry out long sequences of keystrokes all at once, saving you a great deal of time and energy.

In this appendix we will show you some simple TI-82 (TI-83) programs and how to enter and use them. In the textbook, there are some other programs which you will find useful in solving problems.

Correcting Mistakes

When you enter a program you will almost surely make some keying mistakes. You can use the arrow keys to back up and key over any mistakes. To insert something new, rather than keying over what is already there, give the insert command (INS above the ⌷ DEL ⌷ key) by keying

⌷ 2nd ⌷   ⌷ DEL ⌷

Use the ⌷ DEL ⌷ key to delete the current character.

Entering Programs

To enter a program, give the ⌷ NEW ⌷ command under the ⌷ PRGM ⌷ menu by keying

⌷ PRGM ⌷   ⌷ ◁ ⌷   ⌷ ENTER ⌷

Your calculator should look like Display C.1. You are now in program writing mode. Whatever you key in will be stored in the program you are creating, rather than being executed directly. To get out of program writing mode, give the QUIT command.

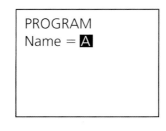

Display C.1

Next you need to name your program so that you can use it. We will start with a very short and not very useful program just to test your ability to enter a program and run it. Give your program the name ADD. Normally, to enter the capital letters that are above and to the right of some of the keys you must press the ALPHA key first. When naming a program, however, the calculator goes into ALPHA mode automatically. This means you *don't* have to press the ALPHA key when entering the letters in the program name. Key in

Your calculator screen should now look like Display C.2

PROGRAM: ADD
: ■

Display C.2

We now need to enter the actual program commands. Our test program will ask for two numbers and then add them. Each line of the program begins with a colon. At the end of each line of the program press ENTER . The first line of our program asks for the first of the two numbers it will add. The two numbers that we tell the calculator to add are called the *input* for the program. The Input command is the first item under the I/O section of the PRGM menu. We will store the input in memory A. Key in

The TI screen should now look like Display C.3.

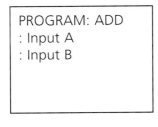

```
PROGRAM: ADD
: Input A
: ■
```

Display C.3

The next line in the program asks for another number and stores it in memory location B. Enter the second line now, based on the way you entered the first line. Your screen should now look like Display C.4.

```
PROGRAM: ADD
: Input A
: Input B
```

Display C.4

The third line adds the numbers stored in memories A and B and stores the result in memory C. Key in

The new screen is in Display C.5.

```
PROGRAM: ADD
: Input A
: Input B
: A + B→C
: ■
```

Display C.5

We finish our program with a statement which displays the result of adding the two numbers (now stored in memory C). The display command Disp is the third item under the I/O section of the PRGM menu.

Appendix C: Programming the TI-82 (TI-83)

Key in

PRGM ▷ 3 ALPHA C

The resulting screen is in Display C.6

```
PROGRAM: ADD
: Input A
: Input B
:A + B → C
:Disp C
:■
```

Display C.6

We are done writing the program! To quit programming mode use the QUIT key (above the MODE key). The program is automatically saved. Key in

2nd QUIT

You are now back to the Home Screen, where you started.

Running Programs

To run the program we just keyed in, we go to the EXEC section of the PRGM menu, key in the number of the program we want to run, and then press ENTER . We will assume that the program named ADD that we just entered is program number 1. Key in

PRGM 1 ENTER

If you entered the program correctly, a question mark appears asking for input. If there is an error, look at the next section on editing programs. This question mark is produced by the first line of your program. You are being requested to type in the first of two numbers, which will then be added by the program. Let's suppose that we want to add the numbers 4 and 5. Press 4 then press ENTER . A second question mark appears asking for the second number. Press 5 and press ENTER again. The result, 9, should appear. The screen now looks like Display C.7.

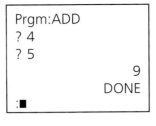

Display C.7

To run the program again, just press the [ENTER] key. You don't have to go through the PRGM menu to run the program the second time, as long as no other calculations have been performed in between. Try adding two other numbers to see how this works.

Quitting Programs

If you are in the middle of running a program and you want to stop the program, press [ON] key. To try this out, run the ADD program again, but this time when the first question mark appears, press [ON] . The screen should look like Display C.8.

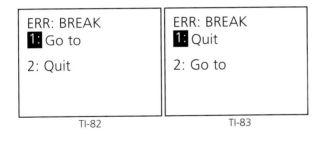

Display C.8

Press [2] to quit the program and return to the Home Screen. Pressing [1] puts you back in program writing mode at the point in the program where you stopped the program.

Editing Programs

If your program doesn't work, or if you just want to make changes to a program, you use the EDIT section of the PRGM menu. Key in

[PRGM] [▷] [1]

This should put you back in the ADD program (assuming it is program 1). Your screen should look just as it did when you left the program writing mode (see Display C.6). Use the arrow keys and the insert INS and delete DEL keys as explained in the Correcting Mistakes section.

To open space for a new line, put the cursor at the beginning of a line, give the insert command INS and then press ENTER . To try this on your ADD program, use the arrow keys to put the cursor at the beginning of the second line of the program and key in

2nd   DEL   ENTER

Your screen should look like Display C.9.

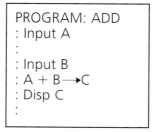

```
PROGRAM: ADD
: Input A
:
: Input B
: A + B⟶C
: Disp C
:
```

Display C.9

The blank line we just created will not affect the program, so we can just give the QUIT command to leave the program writing mode.

A Useful Program

Now that you have some practice with writing, editing and running programs, let's take a look at a program that you might really find useful.

Graphing With Parameters

Suppose that we want to graph the equation of a straight line, say $y = ax + 5$, for several values of a. The constant a is called a *parameter*. First we can enter the expression AX + 5 as expression Y1 under the Y= menu. We can then store numbers in memory A and press GRAPH . The problem is that we only see the graph for one value of A at a time. The following program allows you to easily produce graphs for many values of A and keep all of the graphs on the screen together.

Enter the program shown in Display C.10, using what you learned from the **Entering Programs** section. Name the program PARAMS. Note: DrawF is item 6 under the DRAW menu (above the PRGM key) for the TI-82. Y₁ is item number 1 of the Function sub menu under the Y-**vars** menu (above the DRAW VARS key). For the TI-83, Y₁ is found by keys

VARS ▷ 1 1 ENTER

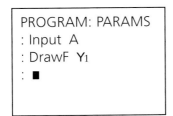

```
PROGRAM: PARAMS
: Input  A
: DrawF  Y1
: ■
```

Display C.10

The program is simple, but saves quite a few keystrokes. You put in a value for A, and then the function is graphed using the DrawF command.

To use this program you must store your function in function memory Y1 and then *turn off* Y1 (put the cursor on the = and press ENTER). Your Y= WINDOW should look like Display C.11. Notice that the = is *not* highlighted, indicating the function is off.

```
Y1 =  AX + 5
Y2 =
Y3 =
Y4 =
Y5 =
Y6 =
Y7 =
Y8 =
```

Display C.11

To set the graph WINDOW to the Standard setting, press the 6 under the ZOOM menu. Your WINDOW settings should appear as in Display C.12.

WINDOW FORMAT	WINDOW FORMAT
Xmin = -10	Xmin = -10
Xmax = 10	Xmax = 10
Xscl = 1	Xscl = 1
Ymin = -10	Ymin = -10
Ymax = 10	Ymax = 10
Yscl = 1	Yscl = 1
	Xres = 1

TI-82   TI-83

Display C.12

Now run the program. Try starting with an A value of 1. Press [1] then [ENTER] in response to the question mark. You should see a graph of the function $y = 1x + 5$. To run the program again, first press the [CLEAR] key (to get back to the Home Screen) then [ENTER]. Try an A with a value of -2 this time. Now both the graphs of Y = AX + B for A = 1 and for A = -2 should be on the screen, as shown in Display C.13.

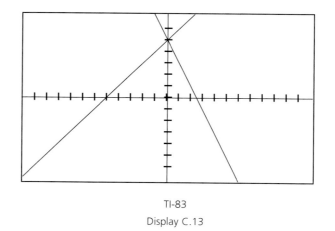

TI-83

Display C.13

If you want to clear the graph screen, use ClrDraw, which is item 1 under the DRAW menu.

You only need to change the function stored in Y₁ to graph any other function with one parameter. For instance, try graphing the function $y = a^x$ for various a values.

GLOSSARY

absolute temperature scale (Kelvin scale) A scale calibrated where water boils at 373° and freezes at 273°. The scale starts at absolute zero and has the same size degrees as the Celsius scale.

absolute value The (nonnegative) distance of a number from 0.

acronym A type of abbreviation for a string of words.

algorithm A step-by-step procedure, often used to solve equations more quickly.

Associative Law of Addition $(a + b) + c = a + (b + c)$

Associative Law of Multiplication $a(b \cdot c) = (a \cdot b)c$

axis One of the reference lines in a coordinate system.

bar graph A picture of data using the length of bars relative to some scale to show the number of times each label occurs.

base The number that is multiplied the number of times indicated by its exponent.

biased (experiment) An experiment or event in which one or several outcomes are more likely to occur than others.

bimodal (data set) A set made up of two distinct data groupings that are relatively far from each other.

boxplot A five-number summary in picture form.

category data Data that can be classified into groups and titled—labels, names, etc.

Celsius temperature scale A scale calibrated where water boils at 100° and freezes at 0°.

coefficient A number or variable that precedes a variable in a product.

commutative A term that means a different order of variables or constants in an expression or an equation.

Commutative Law of Addition $a + b = b + a$

Commutative Law of Multiplication $a \cdot b = b \cdot a$

compound interest Paying interest on interest previously earned.

constant A number, letter, or symbol that represents one value only.

convention An agreement to do something in a particular way.

coordinate Each of the numbers or letters used to identify the position of a point in a coordinate system.

coordinate system A graphical method that uses two or more numbers or letters to locate a position in space. Sometimes called the Cartesian coordinate system after the French Scholar René Descartes who first used this system.

data Factual information.

deviation The difference between a data value and some measure of center, such as the mean.

Distributive Law
 for multiplication over addition
$$a(b + c) = ab + ac$$
$$(a + b)c = ac + bc$$
 for multiplication over subtraction
$$a(b - c) = ab - ac$$
$$(a - b)c = ac - bc$$

dotplot A data display in which each data item is shown as a dot above its value on a number line.

ellipsis The symbol ... used to indicate something is missing or omitted.

empty set The set with no elements.

equation A symbolic statement that two quantities are equal.

event A subset of a sample space.

exponent The small, raised number, or power, that tells how many times a base is multiplied by itself.

exponential decay The description of a function in the form $y = a^x$, where $0 < a < 1$; y decreases as x increases.

exponential growth The description of a function in the form $y = a^x$, where $a > 1$; y increases as x increases.

exponentiation An operation of arithmetic where the base (a number or variable) is raised to an exponent (power).

extrapolation A procedure used to predict values of a variable in an unobserved interval (outside the interval) from observed values inside the interval. Backward extrapolation is when estimate is made to the left of the region of data points; forward extrapolation, or forecasting is when an estimate is made to the right of the region of data points.

Fahrenheit temperature scale A scale calibrated at which water boils at 212° and freezes at 32°.

first differences Obtained by subtracting the previous term from the current term in a series.

first quartile The median of the "first half" of a data set; the median of the data that is less than the median of the entire data set.

five-number summary The minimum, first quartile, median, third quartile, and maximum numbers of a set of numerical data.

forecasting A procedure used to predict future events such as sales, the weather, etc., based on known data.

frequency The number of times a value appears in a data set.

histogram A graphical display of measurement data that uses rectangles to show the frequency of data items in successive numerical intervals of equal size.

inequality An algebraic statement which says that two numbers or expressions representing numbers are *not* equal.

interpolation A procedure used to estimate values of a variable between two known values.

interquartile range The numerical difference between the first and third quartiles of a data set.

Kelvin temperature scale (absolute temperature scale) A scale calibrated where water boils at 373° and freezes at 273°. The scale starts at absolute zero and has the same size degrees as the Celsius scale.

kiloWatt One thousand watts of electrical power.

kiloWatt-hour (kWh) The amount of electrical energy used in one hour.

label data Category data; data that are not numerical.

Laws governing equality If $a = b$, then $a + c = b + c$.
If $a = b$, then $a - c = b - c$.
If $a = b$, then $a \cdot c = b \cdot c$.
If $a = b$ and $c \neq 0$, then.
$$a \div c = b \div c.$$

least-squares line (or **regression line**) A straight line that minimizes the sum of the squares of the vertical distances from the line to the points of a given data set in a coordinate plane.

linear equation An equation that can be put in the form $y = ax + b$ or $x = k$.

linear interpolation A procedure used to estimate values of a variable between two given points in a scatterplot by using the values of the points that lie on a straight line joining the two given points.

mathematical law A statement that is true for all values for which it is defined.

mean The sum of a collection of numerical data divided by the number of data items.

mean absolute deviation The mean (average) of all the distances between the individual data items and the mean of a data set.

measurement data Data that are numbers being used to count or measure things.

median The middle value of a set of numerical data when the data are arranged in size order. If there is an even number of data items, the median is the mean of the two middle data items.

mode the category in a data set that occurs most often.

Ohm's Law Voltage equals resistance times electrical current.

ordered pair Two variables or numbers in which the order matters.

origin The point of intersection of the axes of a coordinate system.

outlier A data item that does not fit in well with the rest of the data; for measurement data, often a number much larger or smaller than most of the other data.

perimeter The measure around the outside shape of a figure.

plot Using a point's coordinates to find its location.

power The exponent that a base is raised to.

quadrant Each of four regions that are formed by the perpendicular axes of a coordinate plane.

quartile One of three numbers (first, second and third quartiles) that separate an ordered set of numerical data into quarters.

radical sign The standard symbol for square root.

range (of a data set) The numerical difference between the largest and smallest data values.

rectangular coordinate system A coordinate system in which the axes and all grid lines intersect at right angles.

recursive algorithm An algorithm that uses the results of one step to obtain the results for the next step.

scatterplot (or **scattergram**) A geometric representation of two-dimensional data in a coordinate plane obtained by plotting the data as points.

scientific notation A form in which numbers are expressed as a single digit number times ten to some integer power.

second quartile The median of a data set.

set-builder notation A way to represent sets by specifying the form of a typical element and the conditions such an element must satisfy. Its structure is "{[form of element] | [conditions]}."

simple interest Interest paid on the original principal only.

slope The ratio $\frac{\text{change in vertical coordinates}}{\text{change in horizontal coordinates}}$ between any two points on the line.

slope-intercept form $y = mx + b$ is the form for a linear equation, where m is the slope and b *is* the y-intercept.

solution (solution set) The set of value(s) for which an equation or inequality is true or the correct answer(s) to a problem.

square (of a number) The product of a number and itself.

square root (of a number) The factor of a number that when squared is the number.

standard deviation The number that is the square root of the variance of a set of data.

stem-and-leaf plot A graphical method of displaying numerical data by grouping items that agree in all but (at most) their final digits.

third quartile The median of the "second half" of a data set.

unit length The distance on each coordinate axis between each unit.

value A number that a variable represents.

variable A letter or other symbol that represents any one of a collection of numbers or things.

variance The mean of the squares of all the deviations of a data set.

Voltage Electric potential expressed in volts.

Watt A unit of electrical power.

x-axis The horizontal axis of a coordinate plane.

x-intercept The x-value of the point where a line or curve intersects the x-axis; i.e., where $y = 0$.

y-axis The vertical axis of a coordinate plane.

y-intercept The point at which a line or curve intersects the y-axis; i.e., where $x = 0$.

Index